SpringerBriefs in Earth Sciences

For further volumes:
http://www.springer.com/series/8897

Vincenzo Pasquale · Massimo Verdoya
Paolo Chiozzi

Geothermics

Heat Flow in the Lithosphere

 Springer

Vincenzo Pasquale
Massimo Verdoya
Paolo Chiozzi
University of Genova
Genova
Italy

ISSN 2191-5369   ISSN 2191-5377   (electronic)
ISBN 978-3-319-02510-0   ISBN 978-3-319-02511-7   (eBook)
DOI 10.1007/978-3-319-02511-7
Springer Cham Heidelberg New York Dordrecht London

Library of Congress Control Number: 2013950043

Printed on acid-free paper

Springer is part of Springer Science+Business Media (www.springer.com)

Preface

Geothermics is a discipline concerned with the study of the heat transport and thermal conditions in the Earth's interior. In its applied aspects, it deals chiefly with the geothermal resource assessment, which implies the determination of how the heat is distributed in the outer layers of the Earth and the evaluation of how much heat could be extracted. In view of the growing interest for such problems, we perceived the need for a comprehensive and modern treatment of the background knowledge of the heat transfer processes in the lithosphere, by including also some techniques to explore the role of water circulation in its uppermost part. After a brief review of the global tectonics and of the structure of the crust and upper mantle, this book introduces the theory of heat conduction as well as the methods for the determination of thermal conductivity and radiogenic heat of rocks. The geothermal flow and the thermal state of the lithosphere and deep interior are then analyzed. The formation, upwelling mechanisms, solidification, and cooling of magmas, which can be a fundamental heat source in many geothermal systems, are also reviewed. Finally, the text focuses on the analytical methods used for gaining information on heat and groundwater flow from the analyses of temperature-depth data. Most of the topics dealt with derives from the research papers screened by peer reviews and published in international journals, that, together with the co-authors of this book, I wrote during several years of work. Data and practical examples are supplied to facilitate the understanding of the different topics. The book is intended for Earth science graduates and researchers. Readers with different backgrounds have to refer to several classic textbooks on geology and geophysics. Finally, I would like to mention with gratitude Mario Pasquale Bossolasco, who many years ago at the University of Genoa kindled my interest in the physics of the Earth.

Genova, June 2013   Vincenzo Pasquale

Contents

Chapter 1
Lithosphere Structure and Dynamics

Abstract This chapter introduces the basic physical features of the lithosphere and underlying asthenosphere, namely body wave velocity, density and elastic properties. It is shown how the lithospheric thickness can be inferred from the elevation of the Earth's surface in regions in isostatic equilibrium. The fundamentals of plate tectonics, the mechanisms which provide the energy dissipated in earthquakes, volcanic eruptions and orogenesis, and the forces which act on the plates are also briefly outlined.

Keywords Seismic waves through the Earth · Earth structure and composition · Lithosphere thickness · Plate tectonics · Driving and resistive forces

1.1 Physical Model

Direct observations have only a minimal role in the knowledge of the structure and composition of the Earth's interior. The study of rocks cropping out at the surface or observed in mines and drillings allows to investigate only the uppermost layers. A major contribution to our knowledge of the internal structure is given by the study of the arrival time and path of elastic waves originated by earthquakes (seismic waves) that allows to infer the presence of discontinuities and the distribution in depth of several physical parameters. The velocity of compressional (v_P) and shear (v_S) body waves can be deduced at any depth from the analysis of travel-time-distance curves of seismic waves and depends on density ρ, bulk modulus or incompressibility K and shear modulus μ

$$v_P = \left(\frac{K + \frac{4}{3}\mu}{\rho} \right)^{1/2} \tag{1.1}$$

V. Pasquale et al., *Geothermics*, SpringerBriefs in Earth Sciences,
DOI: 10.1007/978-3-319-02511-7_1, © The Author(s) 2014

$$v_S = \left(\frac{\mu}{\rho}\right)^{1/2} \tag{1.2}$$

From (1.1) and (1.2), we can infer the Poisson's ratio v, also termed lateral contraction ratio

$$v = \frac{\left(\frac{v_P}{v_S}\right)^2 - 2}{2\left[\left(\frac{v_P}{v_S}\right)^2 - 1\right]} \tag{1.3}$$

and the seismic parameter ϕ as

$$\phi = v_P^2 - \frac{4}{3} v_S^2 \tag{1.4}$$

With good approximation, for the crust $K = 5\mu/3$, therefore

$$\frac{v_P}{v_S} = \sqrt{3} \tag{1.5}$$

and the Poisson's ratio is 0.25 (e.g. see Båth 1979; Bullen and Bolt 1985).

   Equations (1.1) and (1.2) are solved for K, ρ and μ by combining them with a further independent relation. A possible approach assumes that density increases with depth only because of adiabatic compression produced by the overlying material. In a spherical symmetry, by neglecting the centrifugal force, the gravity acceleration g is given by

$$g = \frac{Gm}{r^2} \tag{1.6}$$

where G is the gravitational constant and m is the mass of the sphere of radius r and density ρ

$$m = 4\pi \int_0^r r^2 \rho \, dr \tag{1.7}$$

In hydrostatic equilibrium conditions, the density variation as a function of r and pressure p from the Earth's center is

$$\frac{d\rho}{dr} = \frac{d\rho}{dp}\frac{dp}{dr} = \frac{d\rho}{dp}(-g\rho) \tag{1.8}$$

Since the seismic parameter ϕ is related to density in the form

$$\phi = \frac{K}{\rho} \tag{1.9}$$

it is possible to write that the change in density with depth z is

$$\frac{d\rho}{dz} = \frac{d\rho}{dp} \frac{G m}{r^2} \frac{K}{\phi} \qquad (1.10)$$

Therefore, for an infinitesimal variation of pressure $dp = K d\rho/\rho$, (1.10) becomes

$$\frac{d\rho}{dz} = \frac{G m \rho}{r^2 \phi} = \frac{g \rho}{\phi} \qquad (1.11)$$

This is the classic Adams-Williamson equation, which after integration allows the calculation of the distribution of ρ with z from $\phi(z)$. A better approximation is obtained with the expression

$$\frac{d\rho}{dz} = \frac{g \rho}{\phi} - \alpha \rho \tau \qquad (1.12)$$

where α is the expansion coefficient and τ the difference between the normal thermal gradient and the adiabatic gradient. The second term of (1.12) takes into account the nonadiabaticity.

Values of the observed body wave velocity and the calculated Poisson's ratio, seismic parameter, density, elastic moduli, pressure and gravity of the outer shell of the Earth are given in Table 1.1. The outer shell is divided into two major layers, the lithosphere (literally, "rock" sphere) and the underlying asthenosphere ("weak" sphere). The lithosphere, which includes the crust and a small portion of the mantle, is divided into plates that horizontally move.

1.2 Discontinuities and Composition

The crust is separated from the mantle by a seismic discontinuity, i.e. a sharp change in elastic wave velocity (Table 1.1). This discontinuity, called Moho, is present worldwide and is rather shallow beneath the oceanic basins, where the average crustal thickness is about 8 km. Besides a sediment cover, the oceanic crust is composed by basalts overlying a gabbro layer, i.e. a rock with the same composition of basalt, but with structural differences. In the continents, the Moho depth is generally of 35 km, ranging from 20–30 km in the flat continental areas to 60–70 km beneath the main mountain chains, thus mirroring the topography of the Earth's surface. The upper crust (with an average thickness of 15 km) is composed by a variety of rocks, prevailingly of granodioritic type. On the basis of the compressional wave velocity, it is believed that the lower crust is formed by granulite (a metamorphic rock) or diorite (an intrusive rock with intermediate silica content). Some localities show within the continental crust a well-developed discontinuity, called Conrad, between upper and lower crust.

The sharp change in elastic wave velocity at the Moho (v_P increases from 7.0–7.6 to 8.1 km s^{-1} and v_S from 4.0–4.4 to 4.5 km s^{-1}) is explained with a

Table 1.1 Physical model of the lithosphere and asthenosphere. z is depth, v_P compressional wave velocity, v_S shear wave velocity, v Poisson's ratio, ϕ seismic parameter, ρ density, K bulk modulus, μ shear modulus, p pressure, g gravity (data after Dziewonski and Anderson 1981; Meissner 1986)

Region		z (km)	v_P (km s^{-1})	v_S (km s^{-1})	v	ϕ (km^2 s^{-2})	ρ (kg m^{-3})	K (GPa)	μ (GPa)	p (GPa)	g (m s^{-2})
Lithosphere	Oceanic crust	0	1.45	0.0	0.50	0.0	1020	2.1	0.0	0.00	9.82
		3	1.45	0.0	0.50	0.0	1020	2.1	0.0	0.03	9.82
		3	4.5	2.6	0.25	11.2	2200	24.7	14.9	0.03	9.82
		11	7.0	4.0	0.25	28.0	2900	81.2	48.7	0.22	9.82
	Continental crust	0	5.6	3.2	0.26	17.7	2600	46.0	26.6	0.00	9.82
		15	6.4	3.6	0.27	23.7	2750	65.1	35.5	0.34	9.83
		15	6.6	3.8	0.25	24.3	2800	68.1	40.4	0.34	9.83
		35	7.6	4.4	0.26	33.5	2900	97.1	56.1	0.60	9.84
	Upper mantle	35	8.1	4.5	0.28	38.6	3380	130.5	68.4	0.60	9.84
		80	8.1	4.5	0.28	38.6	3375	130.3	68.3	2.45	9.86
Asthenosphere		80	8.1	4.5	0.28	38.6	3375	130.3	68.3	2.45	9.86
		220	8.0	4.4	0.28	38.2	3360	128.3	65.0	7.11	9.90
		220	8.6	4.6	0.30	45.7	3435	157.1	72.7	7.11	9.91
		400	8.9	4.8	0.29	48.5	3540	171.7	81.6	13.35	9.97

For a liquid ($\mu = 0$) $v = 0.50$

change in chemical composition. The mantle below the crust has an ultrabasic (peridotitic) composition, as magnesium-rich olivine (forsterite) is the most common mineral. Peridotitic rock types include dunite, which is almost entirely composed of olivine, harzburgite and lherzolite, formed by olivine plus orthopyroxene, and olivine plus clinopyroxene/orthopyroxene, respectively, with or without spinel and garnet. The possible composition of the upper mantle can be synthesized in the definition of pyrolite (Ringwood 1975), a theoretical rock inferred from a model envisaging the upper mantle melting to yield basalt and a residual of dunitic composition. In this petrographic model, which is consistent with elastic wave velocity data, the 400 km discontinuity is associated with a phase change from olivine to β-spinel. An alternative model envisages a change from garnet-lherzolite to eclogite (garnet + pyroxene) or olivine-eclogite at a depth of 220 km. Below this depth, the mantle is supposed to be eclogitic (the high-pressure equivalent of basalt).

Deviations from the spherical symmetry in the mantle structure are put into evidence by seismic tomography. Lateral variations of velocity occur throughout the mantle. By analysing the dispersion of surface seismic waves, it was demonstrated that in the uppermost part of the mantle there is an elastic anisotropy. Such anisotropy is of the order of 2–4 % for elastic waves with larger horizontal velocity. This suggests that anisotropic minerals, like olivine, might probably have a preferential orientation.

The lithosphere thickness is on average about 80 km (Table 1.1). The oceanic lithosphere thickness varies from a few kilometers at midocean ridges, where it forms, to more than 100 km in the oldest oceanic basins. The thickness of the continental lithosphere ranges from 30–50 km in the geologically young regions to 250–300 km beneath the continental shields. The lower boundary of the asthenosphere corresponds to that of the upper mantle, which ends at a depth of about 400 km.

In a region in isostatic equilibrium the elevation of the Earth's surface is a measure of the buoyancy of the underlying asthenosphere (Fig. 1.1). The load of a lithosphere column of unit section, height L and density ρ_L equals the weight of the displaced asthenosphere of density ρ_a according to

$$L \rho_L g = (L - H) \rho_a g \tag{1.13}$$

or

$$H \rho_a = L (\rho_a - \rho_L) \tag{1.14}$$

where g is the gravity acceleration and H is the thickness of the lithospheric column above the asthenosphere. The second term of (1.14), multiplied by the gravity acceleration, quantifies the buoyancy exerted on a lithosphere totally submerged by the asthenosphere (Lachenbruch and Morgan 1990).

By indicating elevation with ε, it is possible to write

$$\varepsilon = H - H_o \tag{1.15}$$

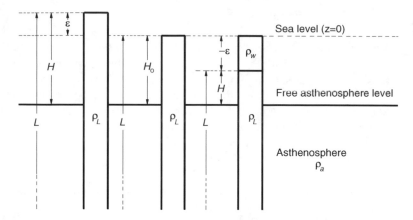

Fig. 1.1 Schematic model of isostatic compensation of the lithosphere. ρ_L, ρ_a and ρ_w densities of the lithosphere, asthenosphere and sea water, respectively; L thickness of the lithosphere; H_o depth of the free asthenospheric level, i.e. without any lithospheric load; ε elevation

where H_o is the distance between the asthenosphere and the sea level. By combining (1.14) and (1.15), a positive elevation is given by

$$\varepsilon = \frac{\rho_a - \rho_L}{\rho_a} L - H_o \qquad\qquad \varepsilon \geq 0 \qquad\qquad (1.16)$$

while a negative elevation (lithosphere beneath sea level) is

$$\varepsilon = \frac{\rho_a}{\rho_a - \rho_w}\left(\frac{\rho_a - \rho_L}{\rho_a} L - H_o\right) \qquad\qquad \varepsilon < 0 \qquad\qquad (1.17)$$

or

$$\varepsilon = \frac{\rho_a - \rho_L}{\rho_a - \rho_w} L - \Delta_o \qquad\qquad \varepsilon < 0 \qquad\qquad (1.18)$$

where

$$\Delta_o = \frac{\rho_a}{\rho_a - \rho_w} H_o \qquad\qquad (1.19)$$

At midocean ridges, where with a good approximation $\rho_a = 3200$, $\rho_L = 2800$, $\rho_w = 1020$ kg m^{-3}, $L = 5.5$ km and $\varepsilon = -2.5$ km, (1.18) and (1.19) give $\Delta_o = 3.5$ km and $H_o = 2.4$ km.

If the role played by the crust and the lithospheric mantle is individually analysed, (1.16) and (1.17) become

$$\begin{cases} \rho_a \varepsilon = (\rho_a - \rho_c)\, h_c + (\rho_a - \rho_{ml})\, h_{ml} - \rho_a H_o & \varepsilon \geq 0 \\[2mm] \dfrac{\rho_a - \rho_w}{\rho_a}\varepsilon = \dfrac{\rho_a - \rho_c}{\rho_a} h_c + \dfrac{\rho_a - \rho_{ml}}{\rho_a} h_{ml} - H_o & \varepsilon < 0 \end{cases} \qquad (1.20)$$

$$\varepsilon = \beta \left(P_c + P_{ml} - H_o \right) \qquad \begin{cases} \beta = 1 & \varepsilon \geq 0 \\[2mm] \beta = \dfrac{\rho_a}{\rho_a - \rho_w} & \varepsilon < 0 \end{cases} \qquad (1.21)$$

where ρ_c and ρ_{ml} are densities and h_c and h_{ml} are thicknesses of the crust and lithospheric mantle, respectively, and β expresses the effect of the water mass on the seafloor subsidence. In (1.21), the contributions to buoyant height of the crust P_c and the lithospheric mantle P_{ml} are

$$\begin{cases} P_c = \dfrac{\rho_a - \rho_c}{\rho_a} h_c \\[3mm] P_{ml} = \dfrac{\rho_a - \rho_{ml}}{\rho_a} h_{ml} \end{cases} \qquad (1.22)$$

P_c is positive because the crust is less dense than the mantle, whereas P_{ml} is negative since the lithospheric mantle is denser than the underlying asthenosphere, which has same composition but is hotter. From (1.21), one obtains

$$P_{ml} = \frac{\varepsilon}{\beta} + H_o - P_c \qquad (1.23)$$

which combined with the second relation of (1.22) allows to obtain the lithospheric thickness L

$$L = h_{ml} + h_c = \frac{\rho_a}{\rho_a - \rho_{ml}} \left(\frac{\varepsilon}{\beta} + H_o - P_c \right) + h_c \qquad (1.24)$$

A lithospheric thickness map of the central Mediterranean obtained with this method is shown in Fig. 1.2. As this area comprises subduction, collision and backarc extension, the lithosphere thickness was calculated where local isostasy applies (Gvirtzman and Nur 2001). Uncertainties about the crust thickness and density, which mainly come from seismic studies, and the lithospheric mantle density may bias of 10–30 km the calculated lithosphere thickness.

1.3 Global Tectonics

At the beginning of the last century, orogenesis was explained with the concept that the Earth was shrinking with time due to cooling; mountain ranges would then be similar the wrinkles in a dry apple, and the horizontal motions of the Earth's crust would be relatively small compared to the vertical displacements. On the basis of the shape similarities between the African and American coast, Alfred Wegener published in 1912 his theory of the continental drift. His idea was that either tidal forces or forces associated with the Earth's rotation caused the present setting of the continents as a result of the breakup of a huge single continent (Pangea) (Fig. 1.3). This dynamic model, based on large horizontal displacements,

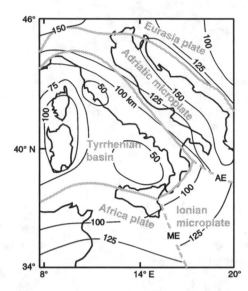

Fig. 1.2 Lithosphere thickness map of the central Mediterranean (based on data in Gvirtzman and Nur 2001; Pasquale 2012). The interactions between the Adriatic microplate and Africa plate are a matter of debate. Like other back-arc basins, the Tyrrhenian area exhibits a subduction slab (Ionian microplate) dipping northwest and corresponding to an active volcanic belt, a thin crust beneath its central part with tholeiitic volcanism, large gravimetric and magnetic anomalies and a high geothermal flow. The Ionian microplate is bordered by two Mesozoic passive margins that might be reactivated along the Apulia (AE) and Malta (ME) escarpments

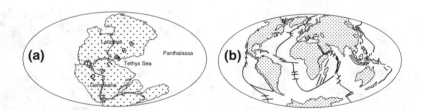

Fig. 1.3 a The old continent of *Pangea* as it may have been 250 Myr ago. The proto-continent broke up forming two supercontinents, *Laurasia* and *Gondwana*; *Panthalassa*, the ocean surrounding Pangea, evolved into the present Pacific Ocean, whereas the Mediterranean is a remnant of the *Tethys Sea*. Since the Proterozoic, about 2500 Myr ago, there have been various cycles of creation and breakup of Pangea. **b** The present-day geography is shown

was seriously criticised by Jeffreys (1929) who showed that these forces were insufficient to drive the continent motion.

In the 1960 and 1970s several pieces of geological evidence led to reconsider Wegner's intuition (Le Pichon 1968; Hallam 1973, for a review). A fundamental role was played by the discovery of midocean ridges, i.e. the submarine volcanic ranges located in the central parts of the oceans. The sediment thickness is nil at midocean ridges and gradually increases towards the continents. This suggests that

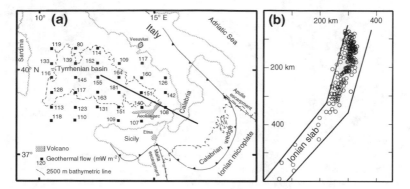

Fig. 1.4 a Geothermal flow and active volcanoes in the Tyrrhenian basin. **b** Distribution of intermediate and deep earthquakes (years 1985–2000) on a cross-section normal to the strike of the Ionian slab (data after Pasquale et al. 2005)

the seafloor is older at the margins and younger in the central part of the ocean basins. Besides paleontological data on sediments, decisive evidence was given by the terrestrial magnetic field and the geothermal flow. Volcanoes are unevenly distributed as they may be either isolated or clustered, mostly in the circum-Pacific area. This distribution generally mirrors that of seismicity, with the exception of few isolated earthquakes related to shallow tectonic motions. Additional evidence is given by the existence of lithosphere slabs, dipping into the mantle, defined by the position of earthquake hypocentres. These zones form compressive structures and occur beneath areas with high geothermal flow and active volcanism (Fig. 1.4).

It is known that the magnetic field undergoes periodic reversals. Lavas, containing ferromagnetic minerals, during their cooling magnetise with the same polarity of the magnetic field. The discovery of alternate magnetic stripes, characterized by anomalously high or low values of magnetic field over the ocean basins, in combination with data on the timing of the magnetic field reversals, led to the idea of sea-floor spreading. The fact that the ocean floor age is much younger (<220 Myr) than the Earth's age was one of the most surprising results of the study of ocean basins (see Cox 1973). These geodynamic processes can be accounted for by the theory of global tectonics. This fundamental concept incorporates several processes, like plate tectonics, sea-floor spreading and mantle convection.

The lithosphere is fragmented into several parts called plates, horizontally moving and transporting ocean and continents over the asthenosphere, which may slowly creep (Fig. 1.5). At midocean ridges, plates move apart, the asthenosphere rises and new oceanic lithosphere forms. Ridges correspond to the diverging margins of the plates, characterised by shallow seismicity, large volumes of basaltic lavas and high geothermal flow.

At convergent margins, where plates subduct, earthquakes can occur up to a depth of 700 km. Volcanism is less widespread than at midocean ridges and has

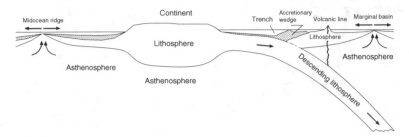

Fig. 1.5 Sketch of the Earth's uppermost layers according to the plate tectonics theory. The crust is welded to the uppermost mantle and together form the lithosphere. The latter is divided into tectonic plates which, rafting on top of the asthenosphere, may move apart or collide

variable chemical features, but is in general of calc-alkaline type, rich in calcium, aluminium, sodium and potassium (alkali). Calc-alkaline magmas have a higher viscosity than basaltic magmas and often produce explosive eruptions. The geothermal flow is high in correspondence of the subducting plate, whereas is low at the trench. Ocean ridges are interrupted by large transform faults, which are characterized by lateral slip without accretion or consumption of lithosphere. Heat originated by friction not always yields high geothermal flow, since most of it is dissipated during the phase and/or state changes. The particular case of transform faults running oblique to the ridge is called "open" fault, along which intense magmatic activity and large geothermal flow can occur.

Seven main tectonic plates of large size have been recognised (the Pacific, North America, South America, Eurasia, Africa, Australian and Antarctic plates) and a number of minor plates (also known as microplates), such as the Nazca, Cocos, Caribbean, Philippine and Arabian plates (Fig. 1.6). Some of them are formed only by oceanic lithosphere (e.g. the Pacific plate), others are instead composed by both continental and oceanic lithosphere (e.g. the Africa plate) or mainly by continental lithosphere (Eurasia plate) (Press and Siever 1974).

One of the most convincing points of the plate tectonics theory lies in the clear explanation that it gives to the problem of orogenesis, i.e. the formation of mountain ranges, such as the Alps and the Himalayas. Part of the shallow crust detaches from the subducting plate and stacks up forming a belt parallel to the margin of the two converging plates. Such a crustal wedge will tend to rise forming a mountain range, because of its density which is lower than that of the mantle.

1.4 Forces Acting on the Plates

Plate motion at a rate of few centimeters per year is related to convective processes in the mantle. The high temperatures in the Earth's interior allow a rise of hot, and consequently, less dense material, whereas elsewhere colder and denser material

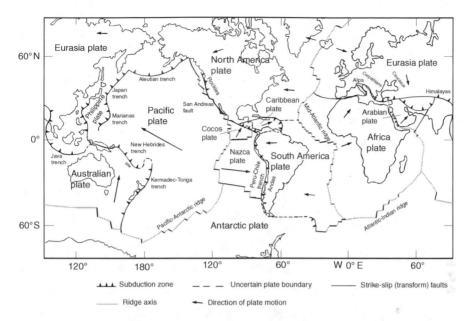

Fig. 1.6 Main tectonic plates and their accreting (spreading ridges), convergent (subduction zones) and transcurrent (transform faults) boundaries. The rate of relative plate motion ranges from 1 to 10 cm per year (modified after Morgan 1972; Bolt 1993)

sinks, thus completing the cycle. At midocean ridges, tectonic plates move apart, asthenosphere flows upward and new oceanic lithosphere forms. When two tectonic plates converge, one sinks beneath the other in the process of subduction. The dip angle of subducting plates ranges from 30 to 70° and depends on the plate velocity and the physical properties of the descending oceanic plate and underlying asthenosphere.

Figure 1.7 shows the main driving and resistive forces acting on the plates. If the plates are moving at a constant velocity, then there must be a force balance: driving forces = resistive forces (Fowler 1990). The driving force F_{RP} (ridge push) acting on the midocean ridges is due to the combination of the rising mantle material and the newly formed plate that subsides and laterally flows at the ridge flanks. The latter contribution to the ridge push is approximately one order of magnitude smaller than the contribution of the rising material. An estimate of F_{RP} per unit length of ridge is given by (Richter and McKenzie 1978)

$$F_{RP} = g\, d\, (\rho_m - \rho_w)\ \left(\frac{L}{3} + \frac{d}{2}\right) \tag{1.25}$$

where d is the ridge elevation above the cooled plate, ρ_m the mantle density at the plate base, ρ_w the water density and L the plate thickness. By assuming $L = 80$ km, $d = 2.5$ km, $\rho_w = 1020$ kg m^{-3}, $\rho_m = 3380$ kg m^{-3} and $g = 9.8$ m s^{-2}, (1.25) gives a force of about 2×10^{12} N m^{-1}.

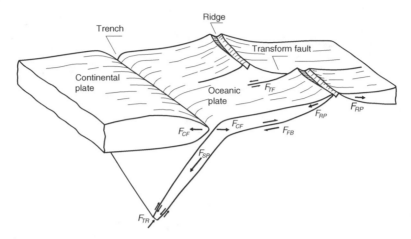

Fig. 1.7 Possible forces acting on plates: F_{RP} ridge push; F_{TF} friction along transform faults; F_{FB} friction at the plate base; F_{SP} slab pull; F_{CF} collision force; F_{TR} mantle resistance to plate descent

The other main driving force, F_{SP} (slab pull), is caused by the subduction of the cool plate at the convergent boundaries. It depends on the density contrast between the plate and the underlying asthenosphere, and can be expressed as

$$F_{SP}(z) = \frac{8\,g\,\alpha\,\rho_m\,T_1\,L^2\,Re}{\pi^4}\left[\exp\left(-\frac{\pi^2 z}{2ReL}\right) - \exp\left(-\frac{\pi^2 b}{2ReL}\right)\right] \qquad (1.26)$$

where z is the depth beneath the base of the plate, α the expansion coefficient, T_1 the mantle temperature, $b + L$ the upper mantle thickness, Re ($= \rho_m c_p vL/(2\,k)$) the thermal Reynolds number, k the thermal conductivity, c_p the specific heat, v the subduction rate. The slab pull decreases with depth and equals zero at $z = b$. At $z = 0$ F_{SP} is about 10^{13} N m^{-1} and thus is larger than the ridge push. Both forces are due to the difference in density, which subsists between the hot and the cold mantle. The hot mantle can only flow upwards to the surface whereas the cold mantle can be transported only downwards. A secondary slab pull within the descending plate is caused by the olivine-spinel phase change, which strengthens the density contrast between the plate and the mantle. This force is about one-half of F_{SP}.

The foregoing driving forces oppose resistive forces that originate locally from friction at the plate base (F_{FB}) (if the mantle flow is faster than the plate velocity, it becomes a driving force) and along transform faults (F_{TF}). Other resistive forces are due to plate collision (F_{CF}) and to the mantle resistance to the plate descent (F_{TR}). The force acting at the plate base is proportional to the plate surface, but it is of the same magnitude of the resistive forces acting on the sinking plate. These resistive forces, which are proportional to the product of mantle viscosity by plate velocity, are about 10^{13} N m^{-1}.

It is extremely complicated to make an estimate of the forces acting on faults. However, the stress drop after strong earthquakes is of the order of 10 MPa. The earthquakes occurring along the ridge axis and transform faults are weak and shallow, and therefore their contribution to the resistive forces can be ignored. It is likely that the total contribution of the resistance forces is equal or smaller than the ridge push. An estimation of the resistive force acting on reverse faults at convergent plate margins is of the order of 10^{12} N m^{-1}. In summary, the main driving force of plate motion is slab pull, and the main resistive force is at the base and along the descending plate.

However, the problem whether the driving mechanism of plate motion is the mantle convection or the forces acting at the plate boundaries, which in turn drag the mantle, is far from being solved. From the analysis of the driving and resistive forces, it is clearly shown that the descending slab pull is the major factor controlling mantle flow. If ridges were simply located above ascending convective columns, then one edge of any plate would correspond to a ridge, whereas the other edge would lie at a subduction zone. This is not the case; for example the Africa and Antarctic plates are almost entirely bordered by ridges, which form where the lithosphere is weaker and the mantle rises to fill the lack of material. Plates which have a subducting portion move at a larger rate.

Therefore, albeit much is still to be cleared about mantle flow and plate drift, the pull force of the sinking slab at convergent margins seems to be the predominant factor both in the thermal and mechanical modelling of mantle flow. In the Achaean, the Earth probably was much warmer than at present time, with temperatures at the base of the lithosphere reaching peaks of 1700 °C, versus the 1250–1350 °C of present-day. The ridge push was 10^{11} N m^{-1} and the slab pull due to the subducting plate 8×10^{12} N m^{-1}. By equalling driving and resistive forces, it is possible to demonstrate that plate drift was occurring above a mantle with a viscosity of 10^{18} Pa s. The drift rate should have been of about 50 cm yr^{-1}, in order to maintain an adequate heat loss into the oceans (Fowler 1990).

In plate tectonics, only the plate peripheral zones are tectonically active. This picture does not take into account intraplate volcanic centers occurring in both oceanic and continental plates. Intraplate volcanism usually yields linear volcanic ridges, along which the more recent products are often located at one end of the ridge. Such a change of age with distance suggests that volcanic ridges track fixed hot spots and represent the culmination of rising plumes of undepleted mantle (Fig. 1.8). Hot spots were first individuated at the Hawaiian Islands, which form a long volcanic chain in the Pacific Ocean, merging in the Emperor Seamount chain. Island volcanism is extinct and rocks become older and older moving away from Island of Hawaii, the only island with active volcanoes, which emit basaltic magmas chemically and isotopically indistinguishable from midocean ridge basalts.

In the Pacific Ocean there are other island chains and seamounts (like Tuamotu) which show the same relation age-distance and a similar geometry. One explanation is that such island chains are originated by a magma source (hot spot) of the deep mantle. An interesting aspect of this hypothesis is that hot spots can be considered, in a first approximation, as a steady reference system, relatively stable compared to mantle convection. The origin of the steady spots and whether or not they can be a clue to account for plate motion are still unresolved problems.

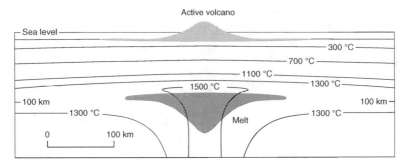

Fig. 1.8 Temperature structure transversally to the Hawaiian Island chain (modified after White and McKenzie 1995). The volume of melt generated beneath the active volcano is modelled with a hot spot having a central temperature of 200 °C greater that the background level

References

Bäth M (1979) Introduction to seismology, 2nd edn. Birkhäuser Verlag, Stuttgart
Bolt BA (1993) Earthquakes. Newly Revised and Expanded. WH Freeman and Company, New York
Bullen KE, Bolt BA (1985) An introduction to the theory of seismology. Cambridge University Press, Cambridge
Cox A (1973) Plate tectonics and geomagnetic reversals. Freeman and Company, San Francisco
Dziewonski AM, Anderson DL (1981) Preliminary reference Earth model. Phys Earth Planet Int 25:297–356
Fowler CMR (1990) The solid earth, an introduction to global geophysics. Cambridge University Press, Cambridge
Gvirtzman Z, Nur A (2001) Residual topography, lithospheric thickness, and sunken slabs in the central Mediterranean. Earth Planet Sci Lett 187:117–130
Hallam A (1973) A revolution in Earth sciences from continental drift to plate tectonics. Clarendon Press, Oxford
Jeffreys H (1929) The Earth, its origin, history and physical constitution, 2nd edn. Cambridge University Press, Cambridge
Lachenbruch AH, Morgan P (1990) Continental extension, magmatism and elevation; formal relations and rules of thumb. Tectonophysics 174:39–62
Le Pichon X (1968) Sea floor spreading and continental drift. J Geophys Res 73:3661–3697
Meissner R (1986) The continental crust: a geophysical approach. Academic Press, Orlando
Morgan WJ (1972) Plate motions and deep mantle convection. Geol Soc Amer Mem 132:7–22
Pasquale V (2012) Geofisica. ECIG—Edizioni Culturali Internazionali Genova, Genova
Pasquale V, Verdoya M, Chiozzi P (2005) Thermal structure of the ionian slab. Pure Appl Geophys 162:967–986
Press F, Siever R (1974) Earth. Freeman WH and Company, San Francisco
Richter FM, McKenzie DP (1978) Simple plate models of mantle convection. J Geophys 44:441–471
Ringwood AE (1975) Composition and petrology of the Earth'mantle. McGraw-Hill, New York
White RS, McKenzie D (1995) Mantle plumes and flood basalts. J Geophys Res 100:17543–17585

Chapter 2
Heat Conduction and Thermal Parameters

Abstract This chapter presents the basic equations for conductive heat transfer and the main thermal parameters of the rocks, in particular the thermal conductivity and radiogenic heat. Also, it outlines the most commonly used techniques for estimating these parameters. Models involving the application of mixing laws for a mineral aggregate are discussed together with techniques for estimating the in situ thermal conductivity and volumetric heat capacity. Finally, methods for determining the radiogenic heat in the crust are introduced.

Keywords Conductive heat flow · Thermal properties · Thermal conductivity measurements · Radiogenic heat · Laboratory and ground gamma-ray spectrometry

2.1 Physical Parameters

It is well known that the heat transfer by conduction prevails in the lithosphere. The equations governing heat conduction include physical parameters such as thermal conductivity, thermal diffusivity and radiogenic heat. In a steady-state thermal regime, i.e. when temperature does not vary with time, thermal conductivity expresses how early heat is transported due to a spatial variation in temperature. Under transient heat conduction, thermal diffusivity describes the rate at which heat flows. Radiogenic heat is produced by naturally occurring radioactive elements, whose decay involves the conversion of mass into energy.

2.1.1 Thermal Properties

Fourier's law states that in isotropic solids the thermal conductivity k is a constant of proportionality between heat-flow density q and temperature gradient ∇T

V. Pasquale et al., *Geothermics*, SpringerBriefs in Earth Sciences, DOI: 10.1007/978-3-319-02511-7_2, © The Author(s) 2014

$$q = -k \, \nabla T \tag{2.1}$$

where the minus sign indicates that heat propagates in the direction of the decreasing temperature. Since q is the amount of energy flowing through a unit area per unit time, k is expressed in W m^{-1} K^{-1}. In the lithosphere, there are three mechanisms which contribute to thermal conductivity: (i) the diffusion of heat by phonon propagation (lattice conductivity k_l), (ii) the transfer of heat through emission and absorption of photons (radiative conductivity k_r) and (iii) the transport of energy by quasiparticles composed of electrons and positive holes (exciton conductibility k_e), whose contribution, however, is negligible in the lithosphere. Thus, k results formed from the sum of the lattice conductivity and radiative conductivity.

At low temperatures, the diffusion of thermal vibrations is the mechanism that essentially contributes to the rock thermal conductivity. In this case, for a single isotropic crystal, at temperatures above the Debye temperature, k_l is given by (Lawson 1957)

$$k_l = \frac{a_i \, K^{3/2}}{3 \, \gamma^2 \, T \, \rho^{1/2}} \tag{2.2}$$

where a_i is the interatomic distance, K is the bulk modulus, T is the absolute temperature, ρ is the density and γ is the dimensionless Grüneisen parameter. The Debye temperature ϑ_D is directly related to the maximum frequency of vibration of the solid v_m as

$$\vartheta_D = \frac{h \, v_m}{k_B} \tag{2.3}$$

where h is the Planck constant ($= 6.626 \times 10^{-34}$ J s) and k_B is the Boltzmann constant ($= 1.381 \times 10^{-23}$ J K^{-1}). For silicate minerals, Horai and Simmons (1970) found an empirical relation between thermal conductivity and Debye temperature

$$\vartheta_D = 61.2 \, k_l + 385 \tag{2.4}$$

with k_l in W m^{-1} K^{-1} and ϑ_D in kelvin. The Grüneisen parameter is weakly dependent on pressure and temperature (for most minerals it varies between 1.0 and 1.5) and expresses the ratio between thermal pressure and thermal energy per unit volume. It is defined as

$$\gamma = \frac{\alpha \, K_S}{\rho \, c_p} = \frac{\alpha \, K_T}{\rho \, c_V} \tag{2.5}$$

where c_p and c_V are the specific heats at constant pressure p and volume V, K_S and K_T are bulk moduli at constant entropy S and temperature T, and α is the thermal expansion coefficient.

In most rocks, k_l ranges from 1 to 7 W m^{-1} K^{-1}, with a few notable exceptions. Experiments confirm the inverse proportionality between k_l and T, and show for temperatures up to about 700 °C the validity of the empirical relation (Schatz and Simmons 1972; Balling 1976)

$$k_l = \frac{1}{a + bT} \tag{2.6}$$

where T is in °C, $a = 0.33$ m K W^{-1} and $b = 0.33 \times 10^{-3}$ m W^{-1} for the upper crust, $a = 0.42$ m K W^{-1} and $b = 0.29 \times 10^{-3}$ m W^{-1} for the lower crust. Thermal conductivity decreases from 3.0 W m^{-1} K^{-1} at the surface to about 2.0 W m^{-1} K^{-1} at 400 °C, and then decreases only very slightly at higher temperatures. In the lithospheric mantle, k_l is 2.0–2.5 W m^{-1}K^{-1}.

The lower crust and mantle temperatures are so high that the contribution of thermal radiation to conductivity must also be considered. At great depth, the heat transport through thermal radiation becomes important, and depends on the mineral opacity ε given by

$$\varepsilon = \frac{\ln (I_o/I)}{x} \tag{2.7}$$

where I_o is the intensity of the incident radiation and I the intensity of the radiation transmitted by a medium of thickness x. If opacity is independent from wavelength, the radiative conductivity k_r is defined by (Clark 1957)

$$k_r = \frac{16\, n^2\, \sigma\, T^3}{3\, \varepsilon} \tag{2.8}$$

where n is the refractive index and σ the Stefan-Boltzmann constant ($= 5.6705 \times 10^{-8}$ W m^{-2} K^{-4}); with $n = 1.74$, a typical value of ferromagnesian silicates, we have

$$k_r = 9.2 \times 10^{-7} \frac{T^3}{\varepsilon} \tag{2.9}$$

If ε were a constant, k_r would increase rapidly with temperature and would then be the dominant mechanism of heat transport. The radiative contribution has been measured on very few minerals. For olivine, by far the best studied, the experimental estimates are highly variable. According to Hasterok (2010), the radiative contribute is negligible at $T < 600$ K and in the lithospheric mantle it can be described by

$$k_r = 0.56 \left[1 + \mathrm{erf} \left(\frac{T - 1150}{370} \right) \right] \tag{2.10}$$

with T in kelvin. Equation (2.10) gives values that significantly differ from previous estimates (Schatz and Simmons 1972; Hofmeister 2005; Hasterok and Chapman 2011). At temperature 1600 K, k_r is 1.1 W m^{-1} K^{-1}.

Finally, we must take account of the effect of pressure on thermal conductivity. By excluding the sedimentary rocks with large porosity (see Sect. 2.4 for the effect on k), the correction on k for igneous and metamorphic rocks at a pressure up to 100 MPa is positive (on average about 10 %). Under a higher pressure, there is a slight increase of k, on average by 0.002 W m^{-1}K^{-1} per 100 MPa, due to the crystal lattice deformation up to the elastic limit (Schloessin and Dvořák 1972).

Another physical property that influences the rate at which heat dissipates through the material is the thermal diffusivity κ, which is defined as

$$\kappa = \frac{k}{\rho\, c} \tag{2.11}$$

where ρ is the density and c the specific heat. For most minerals and rocks, κ is of the order of 10^{-6} m^2 s^{-1}. Diffusivity values of bituminous coal and rock salt are 0.2×10^{-6} m^2 s^{-1} and 3×10^{-6} m^2 s^{-1}, respectively. Grough (1979) derived a diffusivity of 0.33×10^{-6} m^2 s^{-1} for the upper mantle. For nonporous rocks, c ranges from about 0.7 to 1.1 kJ kg^{-1} K^{-1} and, at constant pressure, it is given by

$$c_p = 0.75\left(1 + 6.14 \times 10^{-4}\, T - 1.928 \times 10^4\, T^{-2}\right) \tag{2.12}$$

where T is the absolute temperature. In case of saturated porous rocks, c can be calculated as a weighted average of the matrix and of the pore-filling fluid specific heat. The water specific heat is 4.2 kJ kg^{-1} K^{-1} at room temperature and pressure , and it doubles at 350 °C and 20 MPa. Under high pressure and temperature, the specific heat at constant volume, c_V, is related to c_p as

$$c_p/c_V = 1 + \alpha\gamma T \tag{2.13}$$

in which T is the absolute temperature and γ the Grüneisen parameter.

The ratio between thermal conductivity and the square root of thermal diffusivity defines the thermal inertia

$$\frac{k}{\sqrt{\kappa}} = \sqrt{k\,\rho\,c_V} \tag{2.14}$$

which is a measure of the rock responsiveness to temperature variations. If $k = 2.3$ W m^{-1} K^{-1} and $\kappa = 0.8 \times 10^{-6}$ m^2 s^{-1}, we may determine a thermal inertial of 2.6 kJ m^{-2} K^{-1} s$^{-1/2}$. The higher specific heat the higher thermal inertia. If temperature variations occur with a characteristic thermal time τ, they will propagate to a distance of the order of $(\kappa\tau)^{1/2}$. Similarly, a lapse of time l^2/κ is required for temperature changes to propagate to a distance l.

Table 2.1 Naturally occurring, long-lived isotopes

Isotope	Abundance (wt%)	Decay mechanism	Half-life (years)	Final daughter product	Heat generation (μW kg^{-1})
^{238}U	99.28	$8\alpha + 6\beta^-$ 5.4 10^{-5} % fission	4.47×10^9	^{206}Pb	91.7
^{235}U	0.71	$7\alpha + 4\beta^-$	7.04×10^8	^{207}Pb	575.0
^{232}Th	100	$6\alpha + 4\beta^-$	1.40×10^{10}	^{208}Pb	25.6
^{40}K	0.0118	89.3 % β^- 10.7 % K 0.001 β^+	1.25×10^9	^{40}Ca ^{40}Ar ^{40}Ar	30.0

Half-life = ln2/decay constant. ^{238} U and ^{235} U abundances do not add to 100 % because ^{234} U occurs in the decay series of ^{238} U. Short-lived thorium isotopes, ^{227} Th, ^{228} Th, ^{231} Th and ^{234} Th, also occur in the decay series of ^{238} U, ^{235} U and ^{232} Th

2.1.2 Radiogenic Heat

On a geological time scale, more than 98 % of the heat now being produced in rocks is yielded by the radioactive decay of ^{238}U, ^{232}Th and ^{40}K. These radioelements have half-lives which can be compared to the age of the Earth and are still sufficiently abundant to be an important source of heat. Their abundance, half-lives and energy release are given in Table 2.1.

The nuclei of these radioactive elements are unstable, i.e. they transform into other elements, typically by emitting (sometimes by absorbing) particles. This process, known as radioactive decay, generally results in the emission of alpha or beta particles from the nucleus. It is often also accompanied by the emission of gamma rays of different energies. In the natural uranium, ^{238}U is the most abundant isotope (99.28 %). The other uranium isotopes arc ^{235}U (0.71 %), which has a shorter half-life than ^{238}U, and ^{234}U, which is of negligible importance (0.01 %). ^{238}U undergoes a series of 14 radioactive decays (8 alpha and 6 beta emission) until it becomes ^{206}Pb, a stable nonradioactive element. ^{235}U decays instead through eleven steps (7 alpha and 4 beta emission) and its final product is ^{207}Pb. Natural thorium is primarily formed by ^{232}Th. All the decay products of ^{232}Th have a comparatively short half-life and each nucleus decays through ten steps (6 alpha and 4 beta emission) to reach the final state as the stable isotope ^{208}Pb. ^{40}K is only 0.0118 % of the potassium found in nature. It presents a single-step decay, either transforming in ^{40}Ca (by beta-emission) or ^{40}Ar (by electron capture). The electron capture accounts for only the 10.7 % of decay events, but it produces the 1.46 MeV gamma-ray emission that is used for measuring the level of activity.

The energy radiated during the decay process is converted into heat by absorption. The radiogenic heat, A, can be calculated from the concentration c of uranium, thorium and potassium as

$$A = \rho \sum P A_s c \qquad (2.15)$$

where ρ is the rock density, P the abundance and A_s the rate of heat generation per kilogram of isotope. By using the parameters of Table 2.1, (2.15) becomes (Chiozzi et al. 2002)

$$A = \rho \; (9.51 \; c_U + 2.56 \; c_{Th} + 3.50 \; c_K) \; 10^{-5} \tag{2.16}$$

where A is in μ W m^{-3}, ρ in kg m^{-3}, the uranium and thorium concentrations (c_U and c_{Th}) in ppm (or mg kg^{-1}) and potassium (c_K) in %.

In general, uranium and thorium occur as trace elements, while potassium possibly as a mineralizer in all igneous, metamorphic and sedimentary rocks. Uranium and thorium are enclosed in the crystal lattice of minerals, in inclusions or ionic exchange positions incorporated or built into the mineral structure, and their amount increases with the silica concentration. The soluble part of these lithophile elements, absorbed at the surface of the crystal or found in pore spaces, becomes mobile as soon as water migrates through the rocks. Such a redistribution, especially of uranium, takes place not only during weathering, but also within the crust during magma cooling and in metamorphic processes. Metamorphism may even lead to redistribution of the less soluble part, enclosed in the crystal lattice.

2.2 Heat Conduction Equation

In the previous section, we saw that Fourier's law correlates the heat-flow density with the temperature gradient. By assuming appropriate boundary and initial conditions, this law can be used to determine the temperature distribution in solids, but only if heat is transferred through conduction. The heat-conduction equation can be obtained by considering a small element of volume δV of parallelepiped shape, with edges δx, δy, δz parallel to the coordinate axes (Fig. 2.1). If the amount of heat entering the volume element is smaller than the heat flowing out, in the x axis direction we have

Fig. 2.1 Heat-flow density q_x in a small volume $\delta V (= \delta x \delta y \delta z)$ in the direction of axis x

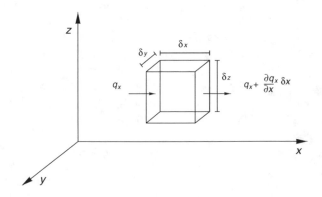

$$\left(q_x + \frac{\partial q_x}{\partial x}\,\delta x\right)\delta y\,\delta z - q_x\,\delta y\,\delta z = \frac{\partial q_x}{\partial x}\,\delta V$$

Similar expressions are obtained for the other two directions, and the amount of heat removed from the volume element is

$$\left(\frac{\partial q_x}{\partial x} + \frac{\partial q_y}{\partial y} + \frac{\partial q_z}{\partial z}\right)\delta V \tag{2.17}$$

By indicating with A the heat production rate per unit of mass (for the lithosphere it correspond to the radiogenic heat) and with $\partial Q/\partial t$ the heat change within δV, we can write

$$\left(\frac{\partial q_x}{\partial x} + \frac{\partial q_y}{\partial y} + \frac{\partial q_z}{\partial z}\right)\delta V = -\frac{\partial Q}{\partial t} + A\,\delta V \tag{2.18}$$

where the minus sign means that the heat loss, greater than the heat internally produced, causes a decrease of heat content in the element. As the variation of heat content with time t is related to the temperature variation as

$$\frac{\partial Q}{\partial t} = c\rho_p \frac{\partial T}{\partial t}\,\delta V \tag{2.19}$$

where ρ and c are the density and the specific heat at constant pressure, (2.18) becomes

$$\left[\frac{\partial\left(-k_x\frac{\partial T}{\partial x}\right)}{\partial x} + \frac{\partial\left(-k_y\frac{\partial T}{\partial y}\right)}{\partial y} + \frac{\partial\left(-k_z\frac{\partial T}{\partial z}\right)}{\partial z}\right]\delta V = -c\rho_p\frac{\partial T}{\partial t}\delta V + A\,\delta V$$

Thus, by dividing by δV and by assuming that thermal conductivity k is constant in the three directions and independent from temperature and pressure, for a homogeneous isotropic solid we have

$$k\,\nabla^2 T = c\rho_p\frac{\partial T}{\partial t} - A \tag{2.20}$$

For steady state conditions $(\partial T/\partial t = 0)$, from (2.20) it is possible to obtain Poisson's equation

$$k\,\nabla^2 T = -A \tag{2.21}$$

which in the absence of internal heat production $(A = 0)$ is reduced to Laplace's equation

$$\nabla^2 T = 0 \tag{2.22}$$

In the case of transient regime and negligible heat production instead, the Fourier equation is obtained

$$\nabla^2 T = \frac{1}{\kappa} \frac{\partial T}{\partial t} \tag{2.23}$$

The assumption of uniform conductivity is only an approximation, and in some cases it is necessary to keep k in the differential operator. The general equation of heat conduction, which gives the temperature within a heterogeneous solid, is then given by

$$\rho\, c_p\, \frac{\partial T}{\partial t} = \nabla k\, \nabla T + k\, \nabla^2 T + A(x, y, z, t) \tag{2.24}$$

which for a steady-state regime becomes

$$\nabla k\, \nabla T + k\, \nabla^2 T + A(x, y, z) = 0 \tag{2.25}$$

By using the vector identity

$$\nabla k = \left(\frac{dk}{dT}\right) \nabla T$$

equation (2.25) can be modified as

$$\frac{dk}{dT}\, (\nabla T)^2 + k\, \nabla^2 T + A(x, y, z) = 0 \tag{2.26}$$

For a semi–infinite half–space defined by $z > 0$, the solution of (2.26) with $k = k_0/(1 + bT)$ and $A = A_o\, \exp(-z/d)$ is (Pasquale 1987)

$$T(z) = \frac{1}{b} \exp \left\{ \ln(1 + b\, T_o) + \frac{q_o b}{k_o}\, z - \frac{A_o b\, d}{k_o} \left[d \exp\left(-\frac{z}{d}\right) + z - d \right] \right\} - \frac{1}{b} \tag{2.27}$$

where k_o is thermal conductivity at room temperature, b and d are constants, T_o, q_o and A_o are temperature, heat-flow density and heat production at depth $z = 0$ (for $b \to 0$, k is constant; for $d \to \infty$, A is constant). For a plane-parallel multilayered model, with A_i and A_{i+1} the heat production at Z_i (upper) and Z_{i+1} (lower) depth of the ith layer, the temperature is given by

$$T(z) = \frac{1}{b_i} \exp \left\{ \ln(1 + b_i\, T_i) + \frac{q_i\, b_i\, (z - z_i)}{k_i} - \frac{A_i\, b_i\, d_i}{k_i} \left[d_i \exp \frac{(z_i - z)}{d_i} + z - z_i - d_i \right] \right\} - \frac{1}{b_i} \tag{2.28}$$

where, b_i and k_i are the parameters (as defined above) of the layer, $d_i = (z_{i+1} - z_i)/\ln(A_i/A_{i+1})$, $q_i = q_{i+1} + A_i d_i\{1 - \exp[(z_i - z_{i+1})/d_i]\}$, T_i and q_i are, respectively, the temperature and heat-flow density at the depth z_i.

2.3 Thermal Conductivity Measurements

Methods for measuring thermal conductivity of soil and rocks may be classified into two categories: steady-state and transient. They both may give absolute or relative values, compared to standards, but in practice the steady-state methods are used in the comparative mode, whereas the transient methods in the absolute mode. Normally, transient methods give thermal diffusivity, however, under certain experimental conditions, they can measure directly thermal conductivity (see e.g. Tye 1969; Parrott and Stuckes 1975; Beck 1988, for a review).

Steady-state measurements on rocks are usually made by means of the divided-bar method, first described by Benfield (1939) and Bullard (1939). There are various modifications to this method, depending on the temperature and pressure needed as well as the rock specimen size (Birch 1950; Beck and Beck 1958; Sass et al. 1971; Blackwell and Spafford 1987). The rock specimen is placed between two blocks of known thermal conductivity. The setup is usually vertical, with the specimen between the hot block at the top and the cold block at the bottom; the heat transfers downwards preventing any convection within the specimen. Measurements are taken when a steady heat flow occurs along the bar.

The most commonly used method for soft rocks is the needle probe, first described by De Vries and Peck (1958) and Von Herzen and Maxwell (1959). It is a transient method based on the theory of an infinite line heat source, embedded within an infinite material. In this configuration, the thermal response is detected by a temperature sensor located on the heat source. For hard rocks, measurements under transient conditions are usually made by means of a plane or a circular heat source, with a temperature sensor at the center (Beardsmore and Cull 2001).

The development of new electronic tools for data acquisition allowed the improvement of some classical methods, such as the line source (see Pribnow and Sass 1995, for a review) and the pulsed line source (Lewis et al. 1993). A further implementation of electronic data aquisition is the scanning of rock specimen surfaces with a focused, mobile and continuously operated, constant heat source, in combination with a temperature sensor (Popov 1983; Popov et al. 1999). Pasquale (1983) revised a method by Joffe and Joffe (1958) which was designed for determining the thermal conductivity of semiconductors, making it suitable for routine measurements on consolidated rocks. A synthesis of some classical methods is presented in the following sections.

2.3.1 Divided Bar

The divided bar is a steady-state method with an accuracy of 2 % (Jessop 1990). It uses a comparative technique in which the temperature drop across a disk of rock

Fig. 2.2 Scheme of the
divided bar method

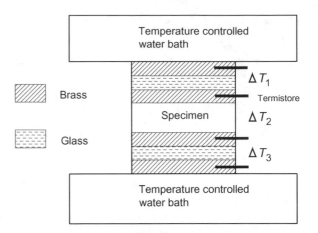

specimen (or of a cylindrical cell containing water-saturated material) is compared
with the temperature drop across two glass disks of known thermal conductivity
(Fig. 2.2). Glass is chosen as the reference material since its thermal resistance is
of the same magnitude as that of the specimen. Since these glass disks cannot be
adapted to allow for the inclusion of temperature sensors, brass disks are inserted
on both sides of each glass disk. The ends of the bar are kept at constant temperature by means of a thermostatic bath. The top bath is held about five degrees
above room temperature and the bottom bath is five degrees below room
temperature.

From each trial, four equilibrium temperature readings are performed. The
relationship for thermal conductivity is based on the principle that the temperature
drop is proportional to thermal resistance. The basic equation is

$$h_g \frac{\Delta T_2}{\Delta T_1 + \Delta T_3} = h_s \frac{k_g}{k_s} + R k_g \qquad (2.29)$$

where k_s and h_s are the thermal conductivity and thickness of the rock specimen, k_g
and h_g are the thermal conductivity and total thickness of the two glass disks, and
R is an estimate of the total thermal resistance associated with the components of
the divided bar

$$R \approx R_2 - (R_1 + R_3) \frac{h_s k_g}{h_g k_s} \qquad (2.30)$$

R_1, R_2 and R_3 are the thermal resistances of the first glass disk, of the rock
specimen and of the second glass disk, respectively, and ΔT_1, ΔT_2, ΔT_3 are the
temperature differences across these elements. These resistances include the various contacts involving the baths, glass disks, rock specimen and brass disks.

If the thermal conductivity of the glass is known, it will be possible to deduce
the thermal conductivity of the rock specimen from the ratio k_g/k_s, obtained as the
linear slope of $h_g \Delta T_2/(\Delta T_1 + \Delta T_3)$ plotted against h_s for a set of specimens of the

same rock with varying thickness. The value of R can be inferred from the intercept. Generally, only one rock specimen is measured and a value of 1.0×10^{-4} K m^2 W^{-1} is assumed for R. A rock specimen of average conductivity and thickness (e.g. a 10 mm disk of sandstone) will give the bar an overall thermal resistance of about 1.0×10^{-2} K m^2 W^{-1}. Since the resistance R is only 1 % of the total thermal resistance across the bar, the approximation of R does not introduce a relevant error into the calculation.

2.3.2 Needle Probe

The needle probe method is based on the theory of an infinite line heat source embedded within an infinite material. The thermal response is detected by a temperature sensor located on the heat source. Typically, the needle probe has a diameter of 1–3 mm and is designed so as to package both the heat source and temperature sensor closely within the probe. A schematic diagram showing the components of a needle probe is shown in Fig. 2.3a.

For an infinite line source with constant power per unit length Q, the thermal conductivity k is given by (Carslaw and Jaeger 1986)

$$k = \frac{Q}{4\pi} \frac{\partial \ln t}{\partial T} \tag{2.31}$$

where t is time and T is temperature. If Q is known, k can be determined directly once $\ln t$ versus T attains linearity. This is usually within 30 s, although a minimum of 200 s is generally required to establish unambiguous gradients. Experimental results show that the probe is able to measure thermal conductivity within a range of 0.1–16 W m^{-1} K^{-1} with an uncertainty lower than 5 %.

This method provides a two-dimensional value for thermal conductivity for a plane perpendicular to the needle axis. The result obtained from a measurement of an anisotropic sample is related to the orientations of the principal axes of the thermal conductivity tensor (k_x, k_y and k_z)

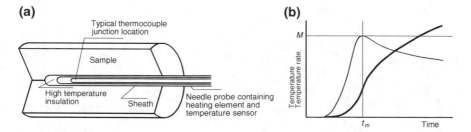

Fig. 2.3 **a** Scheme of the needle probe method. **b** Temperature (*thick line*) and temperature rate (*thin line*) versus time at a distance r from the heat source

$$k = \sqrt{k_x\, k_y \cos^2\gamma + k_x\, k_z \cos^2\beta + k_y\, k_z \cos^2\alpha} \qquad (2.32)$$

where α, β and γ are angles between the line-source axis and the principal axes of thermal conductivity x, y and z, respectively (Popov et al. 1999). Due to layering, many sedimentary rocks are transversely isotropic $(k_x = k_y)$. The thermal conductivity tensor can then be defined by two principal components that are perpendicular $(k_{per} = k_z)$ and parallel $(k_{par} = k_x = k_y)$ to layering. When the needle is parallel to the bedding ($\alpha = 0$ and $\beta = \gamma = 90°$), (2.32) can be simplified into

$$k = \sqrt{k_{par}\, k_{per}} \qquad (2.33)$$

i.e. the measured thermal conductivity is the geometric mean of the parallel and perpendicular components.

As the temperature gradient may induce convection within rock specimen with large porosity, the maximum gradient technique has been developed (Cull 1975). The rate of the temperature increase at a distance r from the line source can be expressed as

$$\frac{\partial T}{\partial t} = \frac{Q}{4\pi k t} \exp\left(-\frac{r^2}{4\kappa t}\right) \qquad (2.34)$$

where κ is the thermal diffusivity. At time t_m, this equation has a maximum value of M (Fig. 2.3b). For $\partial^2 T / \partial t^2 = 0$ we obtain $\kappa = r^2/(4t_m)$, which substituted back into (2.34) gives

$$k = 0.0293\, Q/(M\, t_m) \qquad (2.35)$$

Since M and t_m can be determined and Q is known, thermal conductivity can thus be calculated.

This technique is put into practice using a dual-needle probe with one needle housing the heating wire and the other the temperature sensor at a distance of about 3 mm. Nevertheless, the dual-needle configuration may be unsuitable for measurements in coarse-grained material. For such rock specimen, measurements are better made with a circular heat source (ring) with a temperature sensor at the center. In this case, (2.35) becomes

$$k = 0.0771\, Q/(M_1\, t_{1m}) \qquad (2.36)$$

where $t_{1m} = r^2/(6\kappa)$ and M_1 is the maximum value of the rate of temperature change at the center of the ring of radius $r_1(= 0.50 - 0.75$ cm) expressed as (Somerton and Mossahebi 1967)

$$\frac{\partial T}{\partial t} = \frac{Q\, r}{4\,(\pi\kappa t^3)^{1/2}} \exp\left(-\frac{r^2}{4\kappa t}\right) \qquad (2.37)$$

Q is heat per unit length of wire per unit of time.

2.3.3 Transient Divided Bar

This method is somewhat similar to that developed by Joffe and Joffe (1958) to measure the thermal conductivity of poor conductors. A thermal stack is arranged, consisting of a cylindrical specimen of rock placed between two cylindrical blocks of copper of known thermal capacity (Fig. 2.4a). The upper block acts as a heat source and the lower block, more massive, as a heat sink. The heat flowing through the specimen is equal to the heat absorbed by the sink; in this way thermal conductivity can be found by measuring the changes in temperature at the source and sink.

The experiment begins when the stack of elements reaches the room temperature T_o. The lower block is then cooled by immersing it into a thermostatic bath whose temperature is 10–15 °C lower than T_o. The temperatures of the upper (T_u) and lower (T_l) blocks are recorded by using thermocouples placed as close as possible to the specimen. After a few minutes a uniform temperature gradient $(T_u - T_l)/h$ is established (Fig. 2.4b).

On the base of the Fourier's law and calorimetric law, the amount of heat removed from the upper block, in a given time step Δt, is given by

$$C_u \, \Delta T_u = \frac{k' S}{h} \int_1^2 (T_u - T_l) \, dt = \frac{k' S}{h} \, (\bar{T}_u - \bar{T}_l) \, \Delta t$$

from which

$$k' = \frac{C_u \, \Delta T_u \, h}{S \, (\bar{T}_u - \bar{T}_l) \, \Delta t} \tag{2.38}$$

(a)   **(b)**

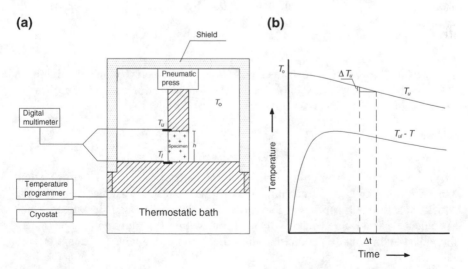

Fig. 2.4 a Scheme of the transient divided bar method. **b** T_u and T_u-T_l curves recorded during the experiment

where

k' is thermal conductivity of the rock specimen,
C_u is the thermal capacity at constant pressure of the upper block,
h is the height of the specimen,
S is the cross-sectional area of the specimen,
ΔT_u is the variation of T_u during time step $\Delta t = t_2 - t_1$,
$(\bar{T}_u - \bar{T}_l)$ is the difference between the average values of T_u and T_l during Δt.

Theoretical analysis of this method (Kaganov 1958; Swann 1959; Drabble and Goldsmith 1961) showed that two additional thermal processes can take place while measuring. The first has to do with the heat coming from the rock specimen, the second with the heat transferred from the surrounding environment to the upper block. Thus, two corrections must be introduced.

The heat flowing from the specimen is taken into account by considering the quantity $(C_u + C_s/3)$ instead of C_u, where C_s is the specimen thermal capacity. In order to estimate the correction for the heat transmitted from the environment, it is necessary to operate under steady-state conditions, when the temperatures T_u and T_l are constant, i.e. about 2 h after the experiment begins. The upper block absorbs heat from the environment, whose temperature T_o is kept constant during the measurement. Under steady-state conditions, the heat Q_s transmitted from the upper block, at constant temperature T_{us}, to the lower block, at constant temperature T_{ls}, will be equal to

$$Q_s = F (T_o - T_{us}) = \frac{k' (T_{us} - T_{ls}) S \Delta t}{h} \tag{2.39}$$

from which it is possible to calculate the quantity F. During the experiment, the relation

$$Q = F (T_o - \bar{T}_u) \tag{2.40}$$

holds, which for each $\bar{T}_u$ allows the calculation of the amount of heat coming from the environment. By taking into account the foregoing corrections, the effective thermal conductivity k is then given by

$$k = \frac{[(C_u + C_s/3) \Delta T_u + F (T_o - \bar{T}_u)] \, h}{(\bar{T}_u - \bar{T}_l) \, S \Delta t} \tag{2.41}$$

A main source of error on thermal conductivity determinations with this method lies in the difficulty of eliminating the thermal contact resistance between the blocks and the rock specimen. Particular care should be taken in preparing the specimen base surfaces, making them flat to within 0.1 mm (no wedge shape), smoothed and mutually parallel to within 0.03 mm throughout (no dome or concavities). A film of silicone paste is smeared on the contact surfaces and a light pressure is exerted on the upper block in order to improve the contact. The thermal resistance of the silicon film about 0.01 mm-thick is equivalent to a rock sheet of a few hundredths of a millimeter. Such a resistance, for an adopted standard height of the specimen (about 2 cm) is less than 2 % of the total thermal resistance of the rock specimen.

If the lateral heat loss is neglected, the time range for which the temperature data should be reliable and a reasonable accuracy achieved falls within the limits of the following inequality (Kaganov 1958)

$$\frac{\ln2\, h^2}{\pi^2 \kappa} \ll t \leq \frac{C_u\, h^2}{C_s \kappa} \tag{2.42}$$

where κ is the rock thermal diffusivity. The order of magnitude of t is of a few minutes. For k ranging from 1 to 10 W m^{-1} K^{-1}, the experimental error is less than 5 % and the reproducibility of about 3 %. This method is unaffected by thermal convection within the specimen as, similarly to the divided bar method, the lower block is kept at a lower temperature. The upper block and the specimen are housed in order to minimize air current effects on the environment temperature. The best performances are obtained when the specimen radius is equal to that of upper block. The specimen height has an effect only if the radius is smaller than the optimal one.

2.3.4 Data Compilation

Thermal conductivity of a variety of water-saturated isotropic sedimentary rocks obtained with the transient divided bar method is given in Table 2.2. In addition to thermal conductivity, porosity and density are listed. Rock porosity is calculated as the ratio of the difference of grain and bulk densities to the grain density. Mass change between dehydrated and water-saturated conditions is accounted for by the influx of water into the pore spaces (this procedure yielded a direct measurement of porosity). The solid volume of argillaceous and evaporitic rocks is inferred with a helium pycnometer. Closed interstices, i.e. the pores that are not connected to the rock surface, are included in the measured volume.

The analyzed rocks were recovered from petroleum exploration wells up to a depth of about seven kilometer and form a broad collection of lithologies. Most of them are clastic and macroscopically isotropic. Clastic rocks consist of framework silicates and carbonate grains in an argillaceous matrix or calcareous cement. Most rocks are marls and silty marls of marine origin, formed by calcium carbonate mud containing a variable amount of clays. The sampled argillaceous sandstones are lithic and feldspatic arenites, primarily composed by cemented sandy sediment, in many cases with a predominance of sand-sized rock fragments and quartz. Chemical/biochemical sediments include carbonatic, evaporitic and siliceous rocks. Evaporitic rocks (anhydrite and gypsum) are fine grained. In anhydrites, gypsum and calcite minerals are also present. Siliceous rocks are represented by radiolarites.

Table 2.2 Thermal conductivity k (water-saturated), porosity and density of isotropic sedimentary rocks (Pasquale et al. 2011)

Rock	Lithotype		n	k (W m⁻¹ K⁻¹)		Porosity (%)		Density (kg m⁻³)	
				Range	Mean	Range	Mean	Range	Mean
Clastic	Marl		19	2.15–3.08	2.77 (0.23)	6.0–37.0	15.1 (8.4)	1787–2530	2278 (240)
	Silty marl		18	2.85–3.66	3.16 (0.26)	2.0–20.0	12.8 (5.5)	2150–2670	2359 (156)
	Calcareous marl		6	1.99–2.37	2.17 (0.13)	22.0–35.0	30.8 (5.0)	1693–2008	1801 (123)
	Argillaceous limestone		3	3.58–3.63	3.60 (0.03)	7.5–12.0	9.3 (2.4)	2477–2588	2520 (80)
	Argillaceous sandstone		6	2.60–3.40	3.00 (0.29)	8.0–25.0	15.1 (6.2)	1990–2560	2330 (222)
	Calcarenite		3	2.18–2.50	2.34 (0.16)	25.0–32.0	29.0 (3.6)	1834–1997	1917 (82)
Chemical-biochemical	Carbonate	Mudstone	5	3.04–3.48	3.30 (0.16)	0.5–6.0	2.7 (2.1)	2550–2695	2630 (59)
		Wackestone	5	3.10–3.20	3.16 (0.04)	3.0–10.0	6.0 (3.0)	2500–2670	2590 (75)
		Packstone	4	3.00–3.45	3.23 (0.18)	3.0–6.0	4.3 (1.3)	2550–2655	2620 (59)
		Grainstone	5	2.95–3.36	3.12 (0.16)	6.5–12.0	8.8 (2.5)	2400–2540	2480 (73)
		Dolostone	5	4.25–5.45	4.60 (0.49)	1.5–7.5	3.8 (2.4)	2630–2800	2735 (73)
	Siliceous	Radiolarite	4	3.16–3.46	3.37 (0.14)	0.5–5.5	2.4 (2.2)	2550–2650	2600 (48)
	Evaporitic	Anhydrite	5	3.15–3.65	3.39 (0.22)	0.5–5.0	2.7 (1.8)	2680–2780	2730 (52)
		Gypsum	5	1.40–1.64	1.54 (0.09)	0.5–7.0	2.4 (2.6)	2260–2400	2350 (61)

In brackets, the standard deviation. n is number of specimens

Table 2.3 Thermal conductivity k and density of igneous and metamorphic rocks (Pasquale et al. 1988)

Lithotype	n	k (W m^{-1} K^{-1})		Density (kg m^{-3})	
		Range	Mean	Range	Mean
Granite	22	2.44–3.49	2.88 (0.26)	2590–2760	2620 (20)
Granodiorite	16	2.24–3.03	2.52 (0.24)	2640–2820	2690 (40)
Tonalite	10	2.06–2.25	2.16 (0.07)	2700–2760	2720 (20)
Syenite	3	2.19–2.34	2.25 (0.08)	2680–2750	2720 (40)
Diorite	14	1.73–2.07	1.89 (0.11)	2740–2940	2840 (60)
Gabbro	12	1.65–2.29	1.94 (0.19)	2800–3060	2940 (80)
Dacite	4	3.56–3.91	3.73 (0.18)	2500–2690	2610 (102)
Anorthosite	4	1.67–1.83	1.76 (0.07)	2660–2810	2730 (60)
Hornblendite	5	2.57–2.79	2.71 (0.08)	3020–3180	3130 (70)
Lherzolite	11	3.31–4.00	3.70 (0.25)	3010–3210	3110 (50)
Harzburgite	3	3.52–3.66	3.60 (0.07)	3070–3110	3090 (20)
Dunite	3	4.04–4.16	4.11 (0.06)	3320–3360	3340 (20)

In brackets, the standard deviation n is number of specimens

The mean thermal conductivity ranges from 1.5 to 4.6 W m^{-1} K^{-1}, corresponding to gypsum and dolostone, respectively. Besides dolostones, larger values of conductivity characterize anhydrites, whereas lower conductivity is typical of calcareous marls. Carbonate rocks and argillaceous sandstones show intermediate values. Porosity varies from about 3 % (radiolarite, gypsum, mudstone and anhydrite) to 30 % (calcarenite and calcareous marl).

Table 2.3 shows thermal conductivity values measured on samples of igneous and metamorphic rocks, macroscopically homogeneous and isotropic. Plutonic rocks are mostly granitoid with a composition ranging from granitic to tonalitic; gabbro-dioritic rocks (syenites, anorthosites and hornblendites) are also listed. Ophiolite rock are mainly peridotites of lherzolitic and harzburgitic type, more or less affected by serpentinization, and, to a smaller extent, of dunitic type, therefore containing almost exclusively olivine. Moreover, there are a few dacitic samples, recovered from deep borehole, belonging to intrasedimentary volcanic bodies. Due to the low porosity of crystalline rocks (mainly below 1.5 %), thermal conductivity is very close to that of rock matrix, and it ranges from 1.8 to 4.1 W m^{-1} K^{-1}.

Data presented in Tables 2.2 and 2.3 are of course additional to several earlier compilations (Birch 1942; Clark 1966; Desai et al. 1974; Kappelmeyer and Häenel 1974; Roy et al. 1981; Čermák and Rybach 1982; Robertson 1988; Zoth and Haenel 1988; Clauser and Huenges 1995). In general thermal conductivity values of igneous and metamorphic rocks are comparable for all datasets, because they depend mainly on mineral composition. For sedimentary rocks, instead thermal conductivity also depends on porosity. Within a particular lithotype, there may be a wide variation in porosity, which may consequently cause big differences in thermal properties. Thus, knowledge of sedimentary rock porosity is essential for an appropriate comparison of datasets.

2.4 Estimates of Thermal Properties

2.4.1 Mixing Models

Besides laboratory techniques, indirect approaches have been developed to estimate thermal conductivity and volumetric heat capacity of isotropic rocks (e.g. Schärli and Rybach 2001; Wang et al. 2006; Abdulagatova et al. 2009). Thermal properties can be inferred from information on volume fraction and conductivity of rock-forming minerals. This approach requires the modeling of the distribution of the various constituents. Literature values of thermal conductivity, density and specific heat of the main rock-forming minerals are summarized in Table 2.4. Thermal parameters for potassium feldspar, plagioclase and sheet silicates correspond to the most abundant minerals, i.e. microcline, oligoclase and the smectite-illite mineral group.

   Provided that the mineral composition and physical properties of minerals and air are known, the volumetric heat capacity of a dry rock $(\rho c)_r$ can be computed as the weighted average of the volumetric heat capacity of the matrix $(\rho c)_m$ and of the air $(\rho c)_a$ in the voids

$$(\rho c)_r = (1 - \phi)(\rho c)_m + \phi (\rho c)_a \tag{2.43}$$

where

$$(\rho c)_m = \sum_{j=1}^{n} v_j \rho_j c_j$$

$$\sum_{j=1}^{n} v_j = 1$$

Table 2.4 Thermal conductivity, density and specific heat of rock-forming minerals, air and water at standard laboratory conditions

Material	Thermal conductivity (W m^{-1} K^{-1})	Density (kg m^{-3})	Specific heat (J kg^{-1} K^{-1})
Quartz-α	7.69[a]	2647[a]	740[b]
Quartz microcrystalline	3.71[a]	2618[a]	735[b]
Plagioclase	1.97[a]	2642[a]	837[b]
K-feldspar	2.40[a]	2562[a]	700[b]
Calcite	3.59[a]	2721[a]	815[b]
Dolomite	5.51[a]	2857[a]	870[b]
Sheet silicates	1.88[c]	2630[c]	832[d]
Anhydrite	4.76[a]	2978[a]	585[b]
Gypsum	1.30[e]	2320[b]	1070[b]
Air	0.026	1.225	1005
Water	0.60	1000	4186

[a] Horai (1971), [b] Waples and Waples (2004a), [c] Brigaud and Vasseur (1989), [d] Hadglu et al. (2007), [e] Clauser and Huenges (1995)

ϕ is porosity, v_j, ρ_j and c_j are the volume fraction, density and specific heat of the jth mineral, respectively, and n is the number of mineral components.

Among the several models involving the application of mixing laws for a mineral aggregate, the model by Hashin and Shtrikman (1962), which was originally proposed to study the magnetic permeability of macroscopically homogeneous and isotropic materials, is currently the most used to calculate thermal conductivity. The matrix conductivity of the rock k_{mo} is given by

$$k_{mo} = \frac{(k_U + k_L)}{2} \tag{2.44}$$

where k_U is the conductivity upper bound defined as

$$k_U = k_{max} + \frac{A_{max}}{1 - a_{max} A_{max}} \tag{2.45}$$

k_{max} is the maximum thermal conductivity of the mineral phases, $a_{max} = (3\,k_{max})^{-1}$ and

$$A_{max} = \sum_{j=1}^{n} \frac{v_j}{\frac{1}{(k_j - k_{max})} + a_{max}} \qquad \text{for } k_j \neq k_{max} \tag{2.46}$$

where k_j is the thermal conductivity of the jth mineral, and v_j and n are as in (2.43). By replacing the minimum thermal conductivity of the mineral phases and the index 'max' with 'min' in (2.45) and (2.46), a similar expression can be obtained for the lower conductivity bound k_L.

In order to determine the bulk conductivity, it is necessary to take into account porosity. According to Zimmerman (1989), pores can be modeled as isolated spheroids and their shape is defined by the aspect ratio, a, i.e. the ratio of the length of the unequal axis to the length of one of the equal axes. Thus, pores have spherical, oblate and prolate shape for $a = 1$, $a < 1$ and $a > 1$, respectively. In extreme cases, i.e. when pores consist of thin cracks, spheroids assume a needle-like, tubular shape $(a \to \infty)$ or thin coin-like shape $(a \to 0)$. For an isotropic rock, an average orientation of the unequal axis of the spheroid with respect to the temperature gradient is considered. If the pores are randomly oriented and distributed spheroids, the computed thermal conductivity k_{HZ} is

$$k_{HZ} = k_{mo} \frac{[(1 - \phi)(1 - r) + r\,\beta\,\phi]}{[(1 - \phi)(1 - r) + \beta\,\phi]} \tag{2.47}$$

where k_{mo} is the matrix thermal conductivity of (2.44), ϕ the porosity and r the ratio of thermal conductivity of the pore-filling water and the matrix thermal conductivity. The parameter β is given by

$$\beta = \frac{(1 - r)}{3} \left[\frac{4}{2 + M(r - 1)} + \frac{1}{1 + (r - 1)(1 - M)} \right]$$

where for $a < 1$

$$M = \frac{2\,\theta - \sin 2\theta}{2\,\tan\theta\,\sin^2\theta}$$

and for $a > 1$

$$M = \frac{1}{\sin^2\theta} - \frac{\cos^2\theta}{2\sin^3\theta}\,\ln\left(\frac{1+\sin\theta}{1-\sin\theta}\right)$$

with $\theta = \cos^{-1}(1/a)$. For $a = 1$, $M = 3(1-r)/(2+r)$.

Another technique to infer thermal conductivity is the geometric mixing model (see e.g. Jessop 1990)

$$k_G = \prod_{j=1}^{n} k_j^{v_j} \qquad (2.48)$$

where symbols are as in (2.46). This method provides a less satisfactory estimate as the difference between computed and measured values is on average of $\pm 5 - 10\,\%$.

2.4.2 In situ Thermal Properties

The assessment of thermal properties at depth is of paramount importance in geothermal studies. For sedimentary rocks, the problem can be approached with a technique based on mineral composition or, alternatively, lithostratigraphic data available from drilling reports (see Pasquale et al. 2008, 2011). The in situ thermal conductivity k_{in} can be obtained with the geometric mixing model

$$k_{in} = k_m^{(1-\phi)}\,k_w^{\phi} \qquad (2.49)$$

where k_m and k_w are the matrix and water thermal conductivity, respectively. Porosity is assumed to decrease with depth z as

$$\phi = \phi_o\,e^{-bz} \qquad (2.50)$$

where b is the compaction factor and ϕ_o is the surface porosity. By expressing depth in kilometers, values adopted for ϕ_o and b are 0.180 and 0.396 km^{-1} in carbonate rocks, 0.298 and 0.461 km^{-1} in marls and silty marls, 0.284 and 0.216 km^{-1} in sandstones and calcarenites, and 0.293 and 0.379 km^{-1} in shales and siltstones, respectively. Carbonate rocks, marls and sandstones are considered as isotropic, whereas in case of clay-rich lithologies (siltstones and shales) thermal anisotropy must be taken into account. In anisotropic rocks, the vertical matrix conductivity, which decreases with depth due to the orientation of the clay and mica platelets during burial, is estimated by using the relation

$$k_m = 2.899 - 0.251z \tag{2.51}$$

The water thermal conductivity k_w is assumed to change with temperature as suggested by Deming and Chapman (1988)

$$k_w = 0.5648 + 1.878 \times 10^{-3}\,T - 7.231 \times 10^{-6}\,T^2 \qquad \text{for } T \le 137\,°\text{C} \tag{2.52}$$

$$k_w = 0.5648 + 1.878 \times 10^{-3}\,T - 7.231 \times 10^{-6}\,T^2 \qquad \text{for } T \le 137\,°\text{C} \tag{2.53}$$

whereas the temperature dependence of the matrix conductivity is evaluated with the expression (Sekiguchi 1984)

$$k_m = 1.8418 + (k_o - 1.8418)\left(\frac{1}{0.002732\ T + 0.7463} - 0.2485\right) \tag{2.54}$$

where k_o is the matrix conductivity at 20 °C. The total uncertainty in thermal conductivity, which takes into account the errors in the corrections for anisotropy, temperature and porosity, is 10 %.

Figure 2.5 shows as an example a profile of k_{in} modeled for two wells. k_{in} is calculated at the middle-point of 20 m intervals. In the uppermost kilometers, the compaction effect is larger than that due to the temperature and, for the same lithotype, this causes an increase of conductivity with depth. Both wells show that the maximum values of k_{in} occur in mudstones and sandstones. Horizons of silty shales are present at different depths and exhibit minima of conductivity. In these horizons, due to the presence of thermally anisotropic sheet silicates, conductivity is constant or decreases with depth.

The in situ volumetric heat capacity $(\rho c)_{in}$ can be computed as the weighted average of the volumetric heat capacity of the matrix $(\rho c)_m$ and the volumetric heat capacity of water $(\rho c)_w$ in the voids

$$(\rho c)_{in} = (1 - \phi)\,(\rho c)_m + \phi\,(\rho c)_w \tag{2.55}$$

The specific heat of both water and matrix increases with temperature. Somerton (1992) proposed an expression for the specific heat of pure water as a function of temperature, which is proven valid in the 20–290 °C range

$$c_w = \frac{(4245 - 1.841\ T) \times 10^3}{\rho_w} \tag{2.56}$$

where ρ_w is the density of liquid water, whose temperature dependence is

$$\rho_w = \frac{\rho_{w20}}{1 + (T - 20)\,\alpha_w} \tag{2.57}$$

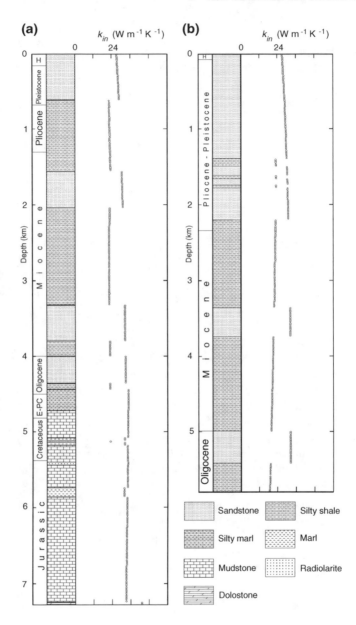

Fig. 2.5 Vertical thermal conductivity as inferred from lithostratigraphic information at Belvedere (**a**) and Sali Vercellese (**b**) wells, Po Basin, Italy (after Pasquale et al. 2012). H = Holocene, E-PC = Eocene–Paleocene

with ρ_{w20} the water density at 20 °C, and α_w the expansion coefficient of water given by

$$\alpha_w = 0.0002115 + 1.32 \times 10^{-6} \, T + 1.09 \times 10^{-8} \, T^2 \qquad (2.58)$$

As long as the pressure is high enough to keep the water in a liquid phase, the volumetric heat capacity of water $(\rho c)_w$, under subsurface (high pressure) conditions, can be estimated with good accuracy from (2.56) to (2.58), without including pressure dependence (Somerton 1992; Waples and Waples 2004b).

Since the thermal expansion coefficient of rocks is very small (about $10^{-5} \, K^{-1}$), in a region of normal geothermal gradient, density can be considered as constant for a wide range of depths. Thus, the temperature dependence of the matrix volumetric heat capacity is controlled by the increase of the specific heat as a function of temperature. The temperature dependence of volumetric heat capacity for any mineral matrix can be computed with the equation (Hantschel and Kauerauf 2009)

$$(\rho c)_m = (\rho c)_{m20} \left(0.953 + 2.29 \times 10^{-3} \, T - 2.835 \times 10^{-6} \, T^2 + 1.191 \times 10^{-9} \, T^3 \right)$$
$$(2.59)$$

where $(\rho c)_{m20}$ is the volumetric heat capacity of the rock matrix at 20 °C.

2.5 Determination of Heat-Producing Elements

Among the different analytical techniques used to determine the concentration of heat-producing elements (see e.g. Rybach 1988, for a brief review), gamma-ray spectrometry is the only one that allows the simultaneous assessement of U, Th and K. Although germanium semiconductor detectors are finding an increasing application in the determination of heat-producing elements, sodium iodide scintillation detectors offer fast and accurate results in ordinary crustal rocks. These detectors are versatile tools for a wide range of geophysical applications, including not only ground radiometric surveys for uranium, vehicle investigations for hydrocarbon and small airborne prospecting, but also laboratory data analysis and core logging services.

2.5.1 Laboratory Techniques

A typical scintillation spectrometer that uses a thallium-activated sodium iodide scintillation detector, coupled to a photomultiplier (PM) tube and a multi-channel (usually 256 or 1024) pulse-height analyzer for storage of the energy spectrum, is shown in Fig. 2.6. The analyzer is usually equipped with a spectrum stabilizer for

Fig. 2.6 A block diagram of
a typical system for gamma-
ray spectrometry

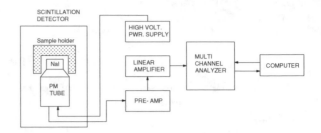

automatic gain shift compensation. The detector is surrounded by a 5–10 cm thick
lead shield, which reduces the amount of background gamma radiation. The
sample preparation requires grinding the rock and encasing it in polyethylene
Marinelli beakers. If a 500 ml beaker is adopted, the investigated rock samples
approximately range in weight from 0.650 to 0.850 kg (Chiozzi et al. 2000a).

The gamma-ray energy spectrum emitted from a rock is the sum of the individual characteristic spectra of each radiogenic component (Fig. 2.7). The total
signal can thus be analyzed to determine the amount of each element. The system
is calibrated by means of three reference materials for K, U and Th activity (IAEA
1987). Usually, the method of the three energy windows described by Rybach
(1971 and 1988) is adopted. Windows are set on the three characteristic energy

Fig. 2.7 Principal gamma-
rays (only energies
>0.100 MeV) during the
decay of ^{238}U and ^{232}Th
series and ^{40}K. Data from
Lederer and Shirley (1978)

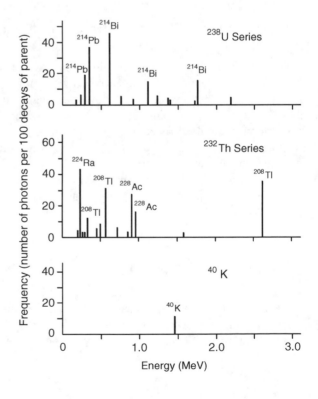

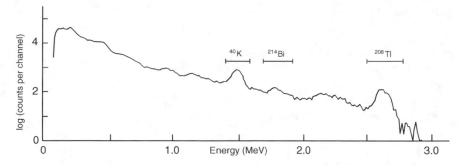

Fig. 2.8 Gamma-ray spectrum obtained for a shoshonitic lava flow and energy windows utilized for data processing (Chiozzi et al. 2003). The counts in each energy channel are represented with a solid line

photo-peaks, i.e. at 1.46, 1.76 and 2.62 MeV, corresponding to ^{40}K, ^{214}Bi (^{238}U series) and ^{208}Tl (^{232}Th series), respectively (Fig. 2.8). The gamma-ray count rate for each spectrum within each energy window depends primarily on (i) the concentration and interference of the radioactive elements; (ii) Compton scattering and the linear attenuation coefficient appropriate to the rock material and the energies of the transmitted gamma rays; (iii) the detector sensitivity and background count rate.

Besides the method of the three energy windows, the radioelement concentrations can be inferred from the total energy spectrum by means of the response matrix method (Matsuda et al. 2002). This technique consists of 22 × 22 elements with unequal energy and pulse height intervals and is applicable to an energy range of 0–3.2 MeV for an isotropic field. Moreover, as the matrix is normalized at the ratio of the count rate to the incident photon fluence rate, one can obtain the flux densities directly from the pulse height distribution if the measured fields are isotropic.

Several factors may affect the accuracy of the determinations of ^{40}K, ^{238}U and ^{232}Th concentrations. Self-absorption due to the density variation in the rock samples seems of negligible importance, and tests on the energy calibration reveal good stability (a maximum drift of 3–4 channels only). Therefore, the main cause of error should lie in the uncertainty of the net count rate. In common rocks the relative standard uncertainty is minimum for potassium (1.2 %), maximum for uranium (3.5 %) and amounts to 2.4 % for thorium. For ultrabasic rocks with exceptionally low radioelement concentration the error increases to 20 %. The experimental standard deviations in carrying out repeated measurements are of the same order of magnitude (Watt and Ramsden 1964; Rybach 1988; Chiozzi et al. 2000a). In a 5400 s time lapse, a typical detection limit is estimated to be of 0.1 ppm for uranium, 0.2 ppm for thorium and 0.02 % for potassium.

2.5.2 Secular Equilibrium

Secular radioactive equilibrium in the decay series (see Faure 1986) is necessary to achieve reliable gamma-ray spectrometry measurements. This condition generally holds in the ^{40}K and for ^{208}Tl in the ^{232}Th series, while problems may arise for ^{214}Bi in the uranium series, where a minimum age of 0.3 Myr is required for ^{226}Ra to be in equilibrium. In young rocks, this problem can be overcome by investigating the gamma-ray spectrum region comprised between 0.03 and 0.10 MeV, instead of the higher energy photo-peak of ^{214}Bi (1.76 MeV) (Ketcham 1996). Chiozzi et al. (2002) tested this approach on young volcanic rocks by comparing results from the NaI(Tl) spectrometer with a germanium detector, and they found out that the use of the low energy window can remarkably reduce the uncertainty in the uranium determination. In this low-energy region, there are a number of gamma rays produced by the decay of ^{234}Th. This radioelement can be safely assumed to be in secular equilibrium with ^{238}U as its half-life is of only 24.1 days, and therefore could have a different activity from post-^{226}Ra daughter products, like ^{214}Bi. As the scintillation detector does not have a sufficient resolution to reliably distinguish individual peaks in this low-energy region, it is necessary to operate with a relatively wide window, ranging from 0.010 to 0.123 MeV.

2.5.3 Ground Spectrometry

Hand-held gamma-ray spectrometry is widely used in mapping surveys aiming at estimating concentrations of U, Th and K of surface rocks (Fig. 2.9). The instrument usually consists of a thallium-activated sodium iodide scintillation detector of about 440 cm^3 together with a photomultiplier tube, a high-voltage supply and a signal preamplifier. The probe is connected to a multichannel analyzer. Like in lab measurements, the determination of these elements is based on the method of the three energy windows (Chiozzi et al. 1998, 2000b).

Field instruments are calibrated by means of standard spectra acquired on three concrete pads enriched in K, U and Th, and with highly unbalanced K/U, Th/U and K/Th ratios. Ideally, calibration pads should simulate a geological source of radiation and should be kept dry, as variation in moisture content may lead to a change in radiation output. Pads are usually of cylindrical shape, with a diameter of 2–3 m and thickness of 0.5 m (IAEA 1989). A fourth, low-radioactivity calibration pad of lead is customarily used to measure the background.

Practically, the calibration procedure consists of measuring the net count rate at each window for each pad, which has finite dimensions and may differ from the recommended size. Grasty et al. (1991) showed that smaller transportable pads are also suitable for calibrating portable gamma-ray spectrometers, provided that a geometrical correction factor, derived from these calibration experiments, is applied to the instrument sensitivity. Furthermore, the detector must be set at a few

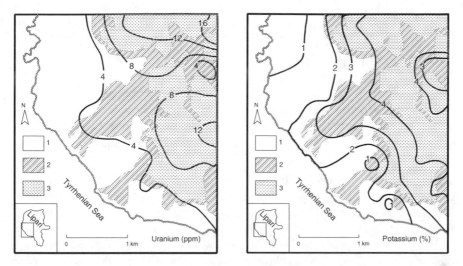

Fig. 2.9 Ground radiometric survey in Lipari, Aeolian volcanic arc, Italy (modified after Chiozzi et al. 2007). Basaltic andesite to high-K andesite (from 223 to 127 kyr ago) (1); high-K andesite (from 127 to 92 kyr ago) (2); rhyolite (from 92 kyr to 1425 yr ago) (3)

centimeters from the surface. In such conditions the count rates are lower than those expected for a 2π infinite geometry. Therefore, the recorded spectra must be corrected for a geometric factor $G = 1-2\ h/d$, where h is the height of the detector to the pad surface and d is the diameter of the standard pad. Usually the determination of the spectrometer sensitivity (with $h = 63$ mm) is based on standard pads of 2 m in diameter (see e.g. Chiozzi et al. 2000b, for details). Under these conditions, the recorded gamma rays are those emitted from a rock volume of about 1 m^3. In underground measurements, such as a in mine gallery, gamma rays emitted from the walls and the vault are also revealed by the detector. This implies an increase in the count rate and, consequently, in concentrations. For a homogenous and isotropic medium and a 4π geometry, measured radioisotope concentrations are expected to double (Bochiolo et al. 2012).

The sampling time required for a measurement depends on the radioactivity of the rock and on the precision required. Lovborg and Mose (1987) derived equations giving the counting time for assays of each radionuclide with a 10 % error at various K, U and Th ratios. In practice, for NaI(Tl) detectors, IAEA (2003) recommends a sampling time of 120 s for highly radioactive rocks and 360 s for low radioactivity outcrops. This implies detection limits of 0.2 and 0.3 ppm for U and Th, respectively, and of 0.03 % for K.

Several factors may affect accuracy in the determination of radioelements. Sources of bias could be inherent to the instrument itself (e.g. drift of energy calibration, calibration of the reference materials, K contamination of the photomultiplier tube) and of geological nature (self-absorption due to variations in rock density, outcrop alteration and geometry). Based on counting statistics, in common rocks the relative combined standard uncertainty of the measured concentrations is of about 3 % for K, 5 % for Th and 8 % for U.

2.5.4 Background Radiation

The net count rate and, consequently, the concentration results depend on the background, i.e. the gamma radiation that does not originate from the rock. Background radiation may be due to the internal radioactivity of the instrument, cosmic radiations—which change with the geomagnetic latitude and elevation—and the atmospheric radon. Therefore, before starting a field radiometric survey, it is important to assess the local background.

Besides on the lead pad, the background can also be estimated by taking measurements on a small boat (preferably of fibreglass) over a lake or sea, at a few hundred meters from a possibly flat shore (Chiozzi et al. 2001; IAEA 2003). Since artificial sources are not always available and, of course, differ from natural conditions, outcrops with energy distributions of gamma spectra similar to those typically emitted from the standards and as close as possible to a 2π geometry can represent an alternative calibration site. Also serpentinitic rocks are a suitable source of background radiation, as their content of K, U and Th is nearly negligible (below the detection limit) (Chiozzi et al. 2000b). The background count rate generally increases with elevation. This is particularly evident in the ^{40}K window. The overall effect is relatively small, but a correction for elevation is often incorporated to improve accuracy (Verdoya et al. 2001).

2.5.5 Alteration Processes

Secular disequilibrium in the U decay series can also be a consequence of alteration processes. On altered rocks, using the low-energy region of the gamma-ray spectrum allows to bypass this problem (see Sect. 2.5.1). Frequently, young volcanics are characterized by hydrothermal activity, which often implies kaolinization. This process yields a group of clay minerals dominated by kaolinite $Al_2(OH)4Si_2O_5$ (in oxides $Al_2O_3 \cdot 2SiO_2 \cdot 2H_2O$), a mineral which belongs to the group of aluminum silicates. Highly-altered argillic zones are often associated with sulphate-acid thermal springs.

The presence of hydrothermally altered zones, where strong remobilization of radioelements is to be expected, conditions the procedure of the radiometric survey, which thus requires both field and laboratory measurements. The detailed study of the spatial distribution and ratios of the radioelements provides a baseline for evaluating the depletion/enrichment processes, which may have affected altered rocks. Field spectrometry can put into evidence low radioactivity levels near the altered zones. The lowest values of K usually occur at kaolin deposits and near hot springs. This is due to the K-depletion process which transforms feldspars and feldspathoids into kaolin.

Kaolinisation takes place in presence of H_2O, CO_2 and high temperatures. In particular, H_2O makes possible, by hydrolysis, the substitution of the alkaline and terrous-alkaline cations in the form of base. At the same time, CO_2, soluble in H_2O, facilitates the transformation of the bases in soluble carbonates that are taken away definitively. Because of these reactions, feldspars and feldspathoids are transformed into colloidal silica and micaceus minerals of the sericite group; the latter can, in turn, lose other silica, and transform by oxidation into clayey minerals, such as illite and kaolinite. When the anhydrous aluminum silicates, which are found in feldspar-rich rocks, are altered by weathering or hydrothermal processes, kaolin is formed (Boulvais et al. 2000; Tourlière et al. 2003; Chiozzi et al. 2007).

Uranium and thorium are commonly fractionated during surface processes due to the oxidation of U into the soluble uranyl ion. The relatively immobile Th tends to concentrate and consequently the Th/U ratio should increase in altered rocks. However, in volcanic areas subject to hydrothermal processes such a ratio shows small variations, giving evidence for a proportional depletion of U and Th.

2.5.6 Radioelement Concentrations and Heat Production

Table 2.5 shows a compilation of concentrations of U, Th and K of a wide set of sedimentary and crystalline rocks, and the radiogenic heat calculated from (2.16). Rock types are sorted according to their depositional environment or to the tectonic regime in which they were formed. Ophiolites, amphibolites, conglomerates and breccias show the lowest concentrations of radionuclides. In general, U and Th are almost always below 1 ppm, and K is lower than 0.7 %. The highest U and Th concentrations are found in metamorphic rocks which also present high concentration of K (migmatites, orthogneisses and micaschists). Th concentration is below the detection limit in lherzolites, prasinites, serpentinites and ophicalcites. In sedimentary and metasedimentary rocks, the average U content is maximum in dolomites, whereas the largest concentrations of Th and K are found in shales and cherts.

Apart from dolomites (which have an anomalous enrichment of U), ophiolitic complex rocks, some limestones, marls, conglomerates and breccias, the Th/U ratio is close to the value expected for a normal continental crust, ranging from 2.3 to 5.4. The estimated radiogenic heat is of course lower in the rocks with lower uranium and thorium concentrations. Excluding amphibolites (0.30 $\mu W\ m^{-3}$), the largest radiogenic heat is found in metamorphic rocks (2.3–3.1 $\mu W\ m^{-3}$). In sedimentary rocks, the radiogenic heat is maximum in shales, cherts, phyllitic schists and calc-schists (2.0–2.6 $\mu W\ m^{-3}$).

Table 2.5 U, Th and K average concentrations, Th/U ratio and radiogenic heat A in rocks of the Ligurian Alps (Italy) (after Chiozzi et al. 2001; Verdoya et al. 2001)

Tectonic/ sedimentary environment	Age	Rock type	n	U ppm	Th ppm	K %	Th/U	A μW m^{-3}
Ophiolitic complex	Middle-late Jurassic	Metagabbro	31	0.3	0.3	0.29	1.00	0.14
		Metabasalt	23	0.4	0.3	0.52	0.75	0.19
		Serpentinite schist	41	0.5	<0.3	0.08	0.60	0.16
	Jurassic	Lherzolite	10	<0.2	<0.3	<0.03	1.50	0.08
		Prasinite	12	0.4	<0.3	0.51	0.75	0.18
	Middle Jurassic	Ophicalcite	10	0.4	<0.3	0.24	0.75	0.15
		Metaophiolite	12	0.6	1.3	0.37	2.17	0.27
Pelagic siliceous sediments	Late Jurassic	Chert	10	3.2	15.2	5.14	4.75	2.20
	Jurassic	Quartz-schist	16	1.2	6.5	1.79	5.42	0.93
Shelf sediments	Middle-late Triassic	Dolomite	15	5.6	0.5	0.17	0.09	1.49
Pelagic sediments	Early Cretaceous	Limestone	14	2.2	5.2	1.25	2.36	1.04
	Lias	Limestone	13	2.6	3.8	1.11	1.42	1.04
Pelitic-pelagic sediments	Early-middle Cretaceous	Phyllitic schist	10	3.0	13.3	3.07	4.43	2.00
	Early Cretaceous	Phyllite	20	2.9	10.6	2.52	3.36	1.73
	Jurassic	Calc-schist	20	3.0	13.3	3.18	4.43	2.01
	Pliocene	Marl	10	2.7	5.5	1.37	2.04	1.21
	Dogger	Calcareous shale	10	2.8	10.1	2.41	3.61	1.66
Flysch	Late Cretaceous	Shale	15	4.1	16.7	3.60	4.07	2.57
	Early Eocene	Marly limestone	10	1.8	4.2	1.15	2.33	0.87
	Middle Cretaceous	Shale	15	2.6	8.5	1.61	3.27	1.42
Molassic deposit	Oligocene	Arenaceous marl	10	2.4	8.8	2.19	3.65	1.45
		Conglomerate	10	1.1	1.6	0.43	1.45	0.42
	Eocene	Breccia	10	0.4	0.3	0.14	0.75	0.14
Metamorphic	Pre-Carboniferous	Orthogneiss	94	4.1	17.6	4.23	4.29	2.65
		Paragneiss	38	3.1	16.8	3.47	5.42	2.30
		Migmatite	6	6.1	16.8	3.83	2.75	3.13
		Amphibolite	31	0.5	1.2	0.69	2.40	0.30
		Micaschist	30	4.1	17.5	3.92	4.27	2.66

n is the number of measurements

Table 2.6 shows an example of the radiogenic heat of rocks belonging to a young volcanic arc and the relative contribution of U, Th and K. The U concentration in rhyolites is remarkably higher than values commonly reported in the literature. The contribution due to U is about one half (44–52 %), whereas that of Th varies from 38 to 44 %. The larger values pertain to trachytic and rhyolitic rocks. The largest contribution yielded by K (about 16 %) is observed in mafic

Table 2.6 Radiogenic heat A of volcanic rocks of the Aeolian arc (Tyrrhenian Sea) and the contribution supplied by each radioelement (after Verdoya et al. 1998a; Chiozzi et al. 2002)

Rock type	A_U %	A_{Th} %	A_K %	A µW m^{-3}
Basalt	43.7 (4.4)	38.4 (7.5)	17.9 (3.0)	0.6 (0.1)
Basaltic andesite	44.9 (4.2)	39.0 (5.7)	16.1 (2.7)	1.1 (0.6)
Andesite	45.7 (4.8)	39.4 (6.6)	14.9 (2.7)	1.3 (0.8)
Basaltic trachyandesite	48.1 (0.3)	43.2 (1.5)	8.6 (1.9)	4.3 (0.8)
Trachyte	52.0 (2.1)	43.0 (3.4)	5.0 (3.5)	7.1 (0.7)
Rhyolite	50.2 (0.7)	44.3 (1.2)	5.4 (0.6)	6.6 (0.9)

Standard deviation in parentheses

rocks. Basalts and andesites present typically low radiogenic heat values (0.6–1.3 µW m^{-3}), whereas trachytic and rhyolitic rocks denote values larger by a factor from six to ten (6.6–7.1 µW m^{-3}). Basaltic trachyandesites are characterized by an intermediate value (4.2 µW m^{-3}).

2.6 Radiogenic Heat in Depth

Experimental results on volcanic rocks show that the radioelement concentration and, consequently, the radiogenic heat are related to magmatic differentiation processes, thus implying a decrease with depth within the crust. The most widely accepted model is the exponential model by Lachenbruch (1970)

$$A(z) = A_o \exp\left(-\frac{z}{D}\right) \qquad (2.60)$$

where A_o (in µW m^{-3}) is the radiogenic heat at the surface and D (in km) is the rate of heat decrease, which realistically explains the differences in the radiogenic heat of surface rocks in a region with a given tectonothermal history (heat-flow province) through the superficial erosion of radioisotopes. Factor D, which ranges from 5 to 15 km and is on average 10 km (Pasquale 1987), derives from the well-known linear relationship

$$q_0 = q_a + A_o D \qquad (2.61)$$

where q_0 is the heat flowing out from the Earth'surface, and q_a is a constant component of heat flow from the mantle. It is widely accepted that the surface radiogenic heat in the continental areas is of the order of 3 µW m^{-3}, as radioactivity measurements of igneous and metamorphic rocks restrict the actual range of radiogenic heat to 2.5–3.5 µW m^{-3}. Pasquale et al. (1990) suggested that radiogenic heat is related to the geological age t according to the empirical relation

$$A_o = 3.2 \exp(-0.3t) \qquad (2.62)$$

where t is in 10^9 years and A_o is in µW m^{-3}.

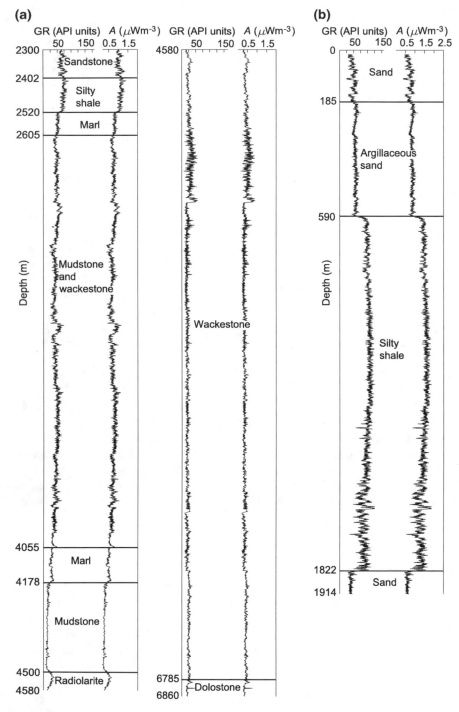

Fig. 2.10 Radiogenic heat *A* derived from GR log of Chiari (**a**) and Carpaneto (**b**) wells, Po Basin (Italy)

Table 2.7 Radiogenic heat A of rocks from GR logs and gamma-ray spectrometry (GRS) (after Pasquale et al. 2012)

Lithotype	A ($\mu W\ m^{-3}$) GR log	A ($\mu W\ m^{-3}$) GRS
Sand	0.74 (0.13)	
Sandstone	1.05 (0.02)	
Siltstone		1.13 (0.12)
Shale/silty shale	1.33 (0.24)	
Marl/silty marl	0.92 (0.14)	1.30 (0.14)
Argillaceous sandstone		1.39 (0.23)
Argillaceous limestone	0.63 (0.14)	
Mudstone/wackestone	0.45 (0.22)	0.34 (0.25)
Dolostone	0.46 (0.32)	
Radiolarite	0.43 (0.09)	
Dacite		0.58 (0.21)
Acid tuff	2.19 (0.15)	
Basaltic tuff	0.47 (0.05)	

Standard deviation in parentheses

Surface radiogenic heat within sediments has not been thoroughly investigated. Information on sediment radiogenic heat can be inferred from petroleum wells. After applying a correction for well diameter, drilling mud density and logging tool eccentricity, gamma-ray logs make possible to estimate A in $\mu W\ m^{-3}$ by means of the relationship (Bucker and Rybach 1996)

$$A = 0.0158\ (GR - 0.8) \tag{2.63}$$

where GR is the log reading in API units. Equation (2.63) may be used for GR values lower than 350 API units and gives A within an acceptable error (<10 %). Figure 2.10 depicts the GR logs and the radiogenic heat profiles for two example wells, whereas the average values of A of several sedimentary rocks are shown in Table 2.7.

At crustal depth, Rybach (1979) and Rybach and Buntebarth (1984) argued that radiogenic heat can be estimated from P-wave velocity v_P, as the gradients in seismic velocity are accompanied by gradients in radiogenic heat. By taking into account the geological age, the relationship between v_P in km s^{-1} and A in $\mu W\ m^{-3}$ is

$$\ln A = B - 2.17 v_P \qquad \text{for } 6.0 < v_P < 8.2 \text{ km s}^{-1} \tag{2.64}$$

where $B = 12.6$ for the Precambrian and 13.7 for the Phanerozoic crust. However, (2.64) takes for granted a simple lithological model of the crust, varying from acid to basic with depth, in agreement with the geochemical constraints. In crustal layers dominated by metamorphic rocks of complex evolution, such a relationship can lead to unreliable results (Kern and Siegesmund 1989).

Table 2.8 Compositional model of the Variscan crust as inferred from the $v_P(z)$ structure and the corresponding radiogenic heat A deduced from petrographical data (after Verdoya et al. 1998b)

Depth range (km)	Lithotype	Percentage of rock type	A (μW m^{-3})
Upper crust			
0–12	Granite-granodiorite	100	3.0
12–18	Granitic gneiss	55	
	Granite-granodiorite	20	1.6
	Tonalitic gneiss	25	
Lower crust			
18–22	Amphibolite	60	0.4
	Mafic granulite	40	
22–30	Mafic garnet granulite	65	0.3
	Amphibolite	35	
30–32	Mafic garnet granulite	100	0.2

Fig. 2.11 Radiogenic heat A versus depth in the Variscan crust based on a compositional model (*grey line*) and the A-v_P relationship (*black line*)

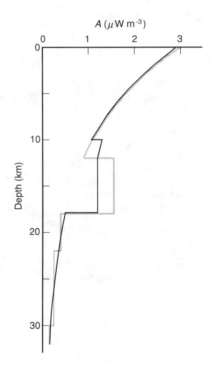

Other possible models of vertical distribution of radiogenic heat rely on both geophysical and petrographical constraints. Verdoya et al. (1998b) proposed an approach that allows the inference of a compositional model by combining crustal seismic profiles with *P*-wave velocity data of rocks from laboratory experiments (Fountain and Christensen 1989). A radiogenic heat value is assigned to each crustal layer on the basis of the compositional model (Table 2.8), and the corresponding $A(z)$ profile is obtained (Fig. 2.11). Equation (2.60) is applied to the granitic-granodioritic component down to a depth of 12 km, for values of $D = 10$ km and surface radiogenic heat of 3.0 μW m^{-3}. Below the uppermost granitic-granodioritic layer, the radiogenic heat is considered uniform with depth. Figure 2.11 also shows a comparison of the compositional model with a model based on (2.62) and (2.64). The radiogenic heat decreases versus depth according to $D = 10$ km. In the lower crust, D was fixed by considering the A value at the top and the bottom of the layer. Below the Moho, A is assumed to be constant (0.01 μW m^{-3}).

References

Abdulagatova Z, Abdulagatov IM, Emirov VN (2009) Effect of temperature and pressure on the thermal conductivity of sandstone. Int J Rock Mech Min Sci 46:1055–1071

Balling NP (1976) Geothermal models of the crust and the uppermost mantle of the Fennoscandian shield in South Norway and the Danish embayment. J Geophys 42:237–256

Beardsmore GR, Cull JP (2001) Crustal heat flow: a guide to measurement and modelling. Cambridge University Press, Cambridge

Beck AE, Beck JM (1958) On the measurement of the thermal conductivity of rocks by observations on a divided bar apparatus. Trans Am Geophys Union 30:1111–1123 (Washington, DC)

Beck AE (1988) Thermal properties. Methods for determining thermal conductivity and thermal diffusivity. In: Haenel R, Rybach L, Stegena L (eds) Handbook of terrestrial heat flow density determination. Kluwer, Dordrecht

Benfield AE (1939) Terrestrial heat in Great Britain. Proc Roy Soc London A 173:428–450

Birch F (1942) Thermal conductivity and diffusivity. In: Birch F, Schaier JF, Spicer HC (eds) Handbook of physical constants, vol 36. Geological Society of America, New York, pp 243–266

Birch F (1950) Flow of heat in the front range Colorado. Bull Geol Soc Am 61:567–630

Blackwell DD, Spafford RE (1987) Experimental methods in continental heat flow. In: Sammis CG, Henyey TL (eds) Methods of experimental physics. Academic Press, Orlando, Florida

Bochiolo M, Verdoya M, Chiozzi P, Pasquale V (2012) Radiometric surveying for the assessment of radiation dose and radon specific exhalation in underground environment. J Appl Geophys 83:100–106

Boulvais P, Vallet JM, Estéoule-Choux J, Fourcade S, Martineau F (2000) Origin of kaolinization in Brittany (NW France) with emphasis on deposits over granite: stable isotopes (O, H) constraints. Chem Geol 168:211–223

Brigaud F, Vasseur G (1989) Mineralogy, porosity and fluid control on thermal conductivity of sedimentary rocks. Geophys J 98:525–542

Bucher C, Rybach L (1996) A simple method to determine heat production from gamma-ray logs. Mar Petrol Geol 13:313–315

Bullard EC (1939) Heat flow in South Africa. Pro Roy Soc London A 173:474–502

Carslaw HS, Jaeger JC (1986) Conduction of heat in solids, 2nd edn. Clarendon Press, Oxford

Čermák V, Rybach L (1982) Thermal conductivity and specific heat of mineral and rocks. In: Angenheister G (ed) Landolt-Börnestein: numerical data and functional relationships in science and technology. Springer, Berlin, pp 305–343

Chiozzi P, Pasquale V, Verdoya M (1998) Ground radiometric survey of U, Th and K on the Lipari Island, Italy. J Appl Geophys 38:209–217

Chiozzi P, De Felice P, Fazio A, Pasquale V, Verdoya M (2000a) Laboratory application of NaI(Tl) gamma-ray spectrometry to studies of natural radioactivity in geophysics. Appl Radiat Isot 53:127–132

Chiozzi P, Pasquale V, Verdoya M, De Felice P (2000b) Practical applicability of field gamma-ray scintillation spectrometry in geophysical surveys. Appl Radiat Isot 53:127–132

Chiozzi P, Pasquale V, Verdoya M (2001) Naturally occurring radioactivity at the Alps-Apennines transition. Radiat Meas 35:147–154

Chiozzi P, Pasquale V, Verdoya M (2002) Heat from radioactive elements in young volcanics by gamma (Ray spectrometry. J Volcan Geoth Res 119:205–214)

Chiozzi P, Pasquale V, Verdoya M, Minato S (2003) Gamma-ray activity in the volcanic islands of the Southern Tyrrhenian Sea. J Environ Radioact 67:235–246

Chiozzi P, Pasquale V, Verdoya M (2007) Radiometric survey for exploration of hydrothermal alteration in a volcanic area. J. Geochem Explor 93:13–20

Clark SP (1957) Radiative transfer in the Earth's mantle. Trans Am Geophys Union 38:931–938

Clark SP (1966) Thermal conductivity. In: Clark SP (ed) Hanbook of physical constants, vol 97. Geological Society of America Memoir, New York, pp 459–482

Clauser C, Huenges E (1995) Thermal conductivity of rocks and minerals. In: Ahrens TJ (ed) Rock physics and phase relations: a handbook of physical constants. American Geophysical Union, Washington

Cull JP (1975) The pressure and temperature dependence of thermal conductivity within the Earth. PhD Thesis, Oxford University, Great Britain

De Vries DA, Peck AJ (1958) On the cylindrical probe method of measuring thermal conductivity with special reference to soils. Aust J Phys 11:255–271

Deming D, Chapman DS (1988) Heat flow in the Utah-Wyoming thrust belt from analysis of bottom-hole temperature data measured in oil and gas wells. J Geophys Res 93:13657–13672

Desai PD, Navarro RA, Hasan SE, Ho CY, DeWitt DP, West TR (1974) Thermophysical properties of selected rocks. Centre for Information and Numerical Data Analysis and Synthesis, Purdue University, West Lafayette, Indiana

Drabble JR, Goldsmith HJ (1961) Thermal conduction in semiconductors. Pergamon Press, New York

Fountain DM, Christensen NI (1989) Composition of the continental crust: A review, in geophysical framework of the continental United States. Geol Soc Am Memoir 172:711–742

Grasty RL, Holman PB, Blanchard YB (1991) Transportable calibration pads for ground and airborne gamma-ray spectrometers. Energy, Mines, and Resources, Ottawa

Grough ST (1979) Geoid anomalies across fracture zones and the thickness of the lithosphere. Earth Planet Sci Lett 44:224–230

Hadglu T, Clinton CL, Bean JE (2007) Determination of heat capacity of Yucca Mountain stratigraphic layer. Int J Rock Mech Min Sc 44:1022–1034

Hantschel T, Kauerauf AI (2009) Fundamentals of basin and petroleum systems modelling. Springer, Berlin

Hashin Z, Shtrikman SA (1962) A variational approach to the theory of the effective magnetic permeability of multiphase materials. J Appl Phys 33:3125–3131

Hasterok D (2010) Thermal state of the oceanic and continental lithosphere. PhD Thesis, University of Utah

Hasterok D, Chapman DS (2011) Heat production and geotherms for the continental lithosphere. Earth Planet Sci Lett 307:59–70

Hofmeister A (2005) Dependence of diffusive radiative transfer on grain-size, temperature, and Fe-content: implications for mantle processes. J Geodyn 40:51–72

Horai K (1971) Thermal conductivity of rock-forming minerals. J Geophys Res 76:1278–1308

Horai K, Simmons G (1970) An empirical relationship between thermal conductivity and Debye temperature for silicates. J Geophys Res 75:978–982

IAEA International Atomic Energy Agency (1987) Preparation of gamma-ray spectrometry reference materials RGU-1, RGTh-1 and RGK-1. Technical Reports Series No. 148, Vienna

IAEA International Atomic Energy Agency (1989) Construction and use of calibration facilities for radiometric field equipment, Technical Reports Series No. 309, Vienna

IAEA International Atomic Energy Agency (2003) Guidelines for radioelement mapping using gamma ray spectrometry data. Technical Reports Series No. 1363, Vienna

Joffe AV, Joffe AF (1958) Measurement of the thermal conductivity of semiconductors in the vicinity of room temperature. Soviet Phys Tech Phys 3:2163–2168

Jessop AM (1990) Thermal geophysics. Elsevier, Amsterdam

Kaganov MA (1958) A theoretical analysis of the method of measuring thermal conductivity of semiconductors proposed by A. V. Ioffe. Soviet Phys Tech Phys 3:2169–2172

Kappelmeyer O, Häenel R (1974) Geothermics with special reference of application. Geoexpl Monographs Gebr Borntraeger, Berlin

Kern H, Siegesmund S (1989) A test of the relationship between seismic velocity and heat production for crustal rocks. Earth Planet Sci Lett 92:89–94

Ketcham RA (1996) An improved method for determination of heat production with gamma-ray scintillation spectrometry. Chem Geol 130:175–194

Lachenbruch AH (1970) Crustal temperature and heat production: implications of the linear heat-flow relation. J Geophys Res 75:3291–3300

Lawson AW (1957) On the high temperature heat conductivity of insulators. J Phys Chem Solids 3:155–156

Lederer CM, Shirley VS (1978) Table of isotopes, 7th edn. Wiley, New York

Lewis T, Villinger H, Davis E (1993) Thermal conductivity measurement of rock fragments using a pulsed needle probe. Can J Earth Sci 30:480–485

Lovborg L, Mose E (1987) Counting statistics in radioelement assaying with a portable spectrometer. Geophysics 52:555–563

Matsuda H, Minato S, Pasquale V (2002) Evaluation of accuracy of response matrix method for environmental gamma ray analysis (in Japanese). Radioisotopes 51:42–50

Parrott JE, Stuckes AD (1975) Thermal conductivity of solids. Pion Ltd, London

Pasquale V (1983) Sulla conducibilità termica delle rocce. Convegno del Gruppo Nazionale di Geofisica della Terra Solida, cnr, Roma, pp 765–775

Pasquale V (1987) Possible thermal structure of the eastern part of the central Alps. Nuovo Cimento 10C:129–141

Pasquale V, Casale G, Masella M (1988) Linear relationships between thermophysical properties and cation packing index of rocks. Preliminary results. Convegno del Gruppo Nazionale di Geofisica della Terra Solida, Cnr, Roma, pp 1423–1431

Pasquale V, Cabella C, Verdoya M (1990) Deep temperatures and lithospheric thickness along the european geotraverse. Tectonophysics 176:1–11

Pasquale V, Chiozzi P, Gola G, Verdoya M (2008) Depth-time correction of petroleum bottom-hole temperatures in the Po plain, Italy. Geophysics 73:E187–E196

Pasquale V, Gola G, Chiozzi P, Verdoya M (2011) Thermophysical properties of the Po basin rocks. Geophys J Int 186:69–81

Pasquale V, Chiozzi P, Verdoya M, Gola G (2012) Heat flow in the Western Po basin and the surrounding orogenic belts. Geophys J Int 190:8–22

Popov YA (1983) Theoretical models of the method of determination of the thermal properties of rocks on the basis of movable sources. Geol Prospect 9:97–103 (in Russian)

Popov YA, Pribnow D, Sass JA, Williams CF, Burkhardt H (1999) Characterization of rock thermal conductivity by high-resolution optical scanning. Geothermics 28:253–276

Pribnow D, Sass JH (1995) Determination of thermal conductivity from deep boreholes. J Geophys Res 100:9981–9994

Robertson EC (1988) Thermal properties of rocks. USGS open file report 88-441, US Geol Survey, Reston, Virginia

Roy RF, Beck AE, Touloukian YS (1981) Thermophysical properties of rocks. In: Touloukian YS, Judd WR, Roy RF (eds) Physical properties of rocks and minerals. McGraw-Hill, New York

Rybach L (1971) Radiometric techniques. In: Wainerdi RE, Uken EA (eds) Modern methods of geochemical analysis. Plenum Press, New York

Rybach L (1979) The relationship between seismic velocity and radioactive heat production in crustal rocks: an exponential law. Pure Appl Geophys 117:75–82

Rybach L (1988) Determination of the heat production rate. In: Rybach L, Stegena L, Haenel R (eds) Handbook of terrestrial heat-flow density determination. Kluwer, Dordrecht

Rybach L, Buntebarth G (1984) The variation of heat generation, density and seismic velocity with rock type in the continental lithosphere. Tectonophysics 103:335–344

Sass JH, Lachenbruch AH, Munroe R (1971) Thermal conductivity of rocks from measurements on fragments and its application to heat flow determinations. J Geophys Res 76:2291–3401

Schärli U, Rybach L (2001) Determination of specific heat capacity on rock fragments. Geothermics 30:93–110

Schatz JF, Simmons G (1972) Thermal conductivity of Earth materials at high temperatures. J Geophys Res 77:6966–6983

Schloessin HH, Dvořák Z (1972) Anisotropic lattice thermal conductivity in enstatite as a function of pressure and temperature. Geophys J R Astr Soc 27:499–516

Sekiguchi K (1984) A method for determining terrestrial heat flow in oil basinal areas. Tectonophysics 103:67–79

Somerton WH (1992) Thermal properties and temperature related behaviour of rock/fluid systems. Elsevier, Amsterdam

Somerton WH, Mossahebi M (1967) Ring heat source probe for rapid determination of thermal conductivity of rocks. Rev Sci Instrum 38:1368–1371

Swann FG (1959) Theory of the AF Ioffe method for rapid measurement of the thermal conductivity of solid. J Franklin Inst 267:363–380

Tourlière B, Perrin J, Le Berre P, Pasquet JF (2003) Use of airborne gamma-ray spectrometry for kaolin exploration. J Appl Geophys 53:91–102

Tye RP (1969) Thermal conductivity, vol. 2. Academic Press, London

Verdoya M, Pasquale V, Chiozzi P (1998a) Radioactive heat production of volcanics. In: Proceedings of the international conference "The Earth's thermal field and related research methods". Moscow, Russia, pp 272–276

Verdoya M, Pasquale V, Chiozzi P, Kukkonen IT (1998b) Radiogenic heat production in the Variscan crust: new determinations and distribution models in Corsica (northwestern Mediterranean). Tectonophysics 291:63–75

Verdoya M, Chiozzi P, Pasquale V (2001) Heat-production radionuclides in metamorphic rocks of the Briançonnais-Piedmont zone (Maritime Alps). Eclogae Geol Helv 94:213–219

Von Herzen RP, Maxwell AE (1959) The measurement of thermal conductivity of deep-sea sediments by a needle probe method. J Geophys Res 64:1557–1563

Wang J, Carson JK, North MF, Cleland DJ (2006) A new approach to modelling the effective thermal conductivity of heterogeneous materials. Int J Heat Mass Transfer 49:3075–3083

Waples DW, Waples JS (2004a) A review and evaluation of specific heat capacities of rocks, minerals, and subsurface fluids. Part 1, minerals and nonporous rocks. Nat Resour Res 13:97–122

Waples DW, Waples JS (2004b) A review and evaluation of specific heat capacities of rocks, minerals, and subsurface fluids. Part 2, fluids and porous rocks. Nat Resour Res 13:123–130

Watt DE, Ramsden D (1964) High sensitivity counting techniques. Pergamon Press, London

Zimmerman RW (1989) Thermal conductivity of fluid saturated rocks. J Petrol Sc Eng 3:219–227

Zoth G, Haenel R (1988) Thermal conductivity. Methods for determining thermal conductivity and thermal diffusivity. In: Haenel R, Rybach L, Stegena L (eds) Handbook of terrestrial heat flow density determination. Kluwer, Dordrecht

Faure G (1986) Principle of isotopes geology, 2nd edn. Wiley, New York

Chapter 3
Thermal State

Abstract The underground temperature, the geothermal flow and the rock radiogenic heat provide evidence of the heat transfer processes and of the thermal state of the lithosphere. Changes in surface temperature can cause fluctuations in the temperature of the uppermost crustal layers. The geothermal flow can be envisaged as a combination of the heat generated from the rock radioactive elements, a transient component caused by tectothermal processes, and the heat flowing out from the asthenosphere. By solving the heat conduction equation for a continental or an oceanic plate, the surface temperature and the geothermal flow allow the determination of the temperature-depth distribution in the lithosphere. Finally, an insight into the thermal regime of the deeper interior is given in this chapter.

Keywords Surface temperature · Geothermal flow · Tectonothermal processes · Lithosphere temperature · Mantle convection · Thermal structure of the core

3.1 Surface Temperature

The uppermost layer of the crust is subject to complex time-dependent physical processes, which involve a continuous heat exchange with the atmosphere. The temperature T as a function of time t and depth z can be found by superposing a background temperature profile due to the geothermal flow on a temperature variation $\Delta T(t, z)$ that depends on the time-varying surface temperature, i.e.

$$T(z,t) = T_o + \Gamma z + \Delta T(z,t) \tag{3.1}$$

V. Pasquale et al., *Geothermics*, SpringerBriefs in Earth Sciences,
DOI: 10.1007/978-3-319-02511-7_3, © The Author(s) 2014

where T_o and Γ are the mean surface temperature and the underground steady-state temperature gradient, respectively. In a homogeneous, isotropic and heat source-free underground, $\Delta T(z, t)$ obeys the diffusion equation (see Sect. 2.2)

$$\frac{\partial T}{\partial t} = \kappa \frac{\partial^2 T}{\partial z^2} \tag{3.2}$$

with positive z-axis downwards and $z = 0$ at the surface. For a surface temperature variation of sinusoidal shape $\Delta T_o \sin(\omega t)$, with $\omega = 2\pi/t_o$ (t_o = period) and amplitude ΔT_o, the solution of (3.2) is (Carslaw and Jaeger 1986)

$$\Delta T(z, t) = \Delta T_o \, e^{-\eta z} \sin(\omega t - \eta z) \tag{3.3}$$

which, by derivating, gives the variation of the thermal gradient

$$\Delta \Gamma(z, t) = -\Delta T_o \eta \, e^{-\eta z} \left[\sin(\omega t - \eta z) + \cos(\omega t - \eta z) \right] \tag{3.4}$$

with $\eta = [\omega/(2\kappa)]^{1/2}$.

The wavelength of oscillation and the propagation velocity are $\lambda = 2\pi(2\kappa/\omega)^{1/2}$ and $v = (2\kappa\omega)^{1/2}$, respectively. If $z = \lambda$, (3.3) becomes

$$\Delta T(t) = \Delta T_o e^{-2\pi} \sin 2\pi \left(\frac{t}{t_o} - 1 \right) \tag{3.5}$$

whose maximum value is about $0.002 \, \Delta T_o$. For a frequency of one cycle per day ($\omega = 2\pi/1 \, \text{day}$) and $\kappa = 30 \, \text{m}^2 \, \text{yr}^{-1}$ this value occurs at a depth of about 1 m. The penetration depth of an annual temperature variation is $(365)^{1/2}$ times larger, i.e. 20 m. The temperature within this surface layer is given by

$$T(z, t) = T_o + \Gamma z + \Delta T_o \, e^{-\eta z} \sin(\omega t - \eta z) \tag{3.6}$$

From (3.3) and (3.4) it follows that $\omega t - \eta z_1 = \pi$ and $\omega t - \eta z_2 = \pi/4$ where z_1 is the first depth at which $\Delta T(z_1) = 0$ and z_2 is the first depth at which $\Delta \Gamma(z_2) = 0$.

In principle, any temperature variation may be treated with (3.3) if its frequency components are separated by Fourier analysis and then summed after attenuation and phase shifting. However, long-term surface temperature variations can be depicted by a step function (Fig. 3.1), i.e. a more straightforward solution to the problem. For a temperature change ΔT_1 (negative for a cold period) beginning at time t_2 and ending at time t_1 measured backwards, we have

$$\Delta T(z, t) = \Delta T_1 \left(\text{erf} \frac{z}{2\sqrt{\kappa t_1}} - \text{erf} \frac{z}{2\sqrt{\kappa t_2}} \right) \tag{3.7}$$

which, in terms of thermal gradient, becomes

$$\Delta \Gamma(z, t) = \Delta T_1 \left[\frac{\exp\left(-\frac{z^2}{4\kappa t_1}\right)}{\sqrt{\pi\kappa \, t_1}} - \frac{\exp\left(-\frac{z^2}{4\kappa t_2}\right)}{\sqrt{\pi\kappa t_2}} \right] \tag{3.8}$$

Fig. 3.1 Single step change.
ΔT_1 is temperature change

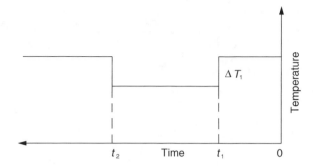

where erf(u), with dimensionless variable $u = z/(4\kappa t)^{1/2}$, is the error function
(Fig. 3.2) defined by

$$\operatorname{erf}(u) = \frac{2}{\sqrt{\pi}} \int_0^u e^{u'^2} \mathrm{d}u' \tag{3.9}$$

This function equals zero for $u = 0$, one for $u \to \infty$ and at $u > 2$ its deviation
from unity is very small. The complementary error function is erfc$(u) = 1 -$
erf(u), erf$(-u) = -$ erf(u), derf$(u)/\mathrm{d}u = 2\pi^{-1/2}e^{-u^2}$.

For a number of step changes, the solution will be given by the sum of each
variation. Figure 3.3 shows an application of (3.8). The climate history of the
western Mediterranean Sea is modeled over the last 100,000 years. Calculations
predict that the maximum deviation in the temperature gradient of the marine
subbottom sediments is of the order of -4 mK m^{-1} near the surface, zero at
1000 m and 2 mK m^{-1} at 2000 m. The temperature change due to the climatic
history reduces and becomes almost negligible at a depth of 6000 m. Such calculations are of paramount importance to correct the temperature gradient and the
geothermal flow for paleoclimatic effect.

Fig. 3.2 Error function for
u < 2

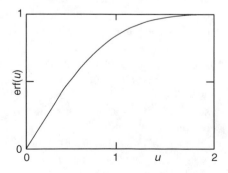

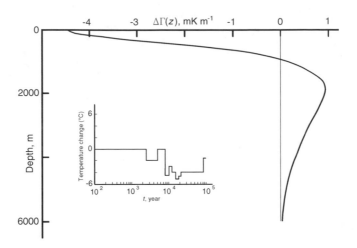

Fig. 3.3 Temperature gradient variation ($\Delta\Gamma$) versus depth due to sea-bottom temperature changes during the last 100,000 years (in *inset*) in the western Mediterranean Sea (data from Pasquale et al. 1996a and 1998). A uniform thermal diffusivity of 30 m^2 yr^{-1} was assumed

3.2 Geothermal Flow

3.2.1 Heat Release

The heat rate per unit area (heat–flow density or heat flux) that vertically flows from the Earth's interior is called geothermal flow (or surface heat flow) q_o and is defined as

$$q_\mathrm{o} = k\Gamma_z \tag{3.10}$$

Conventionally q_o is taken as positive, even if the depth z is positive in the direction of the increasing temperature (downwards). Geothermal flow determinations require separate measurements of thermal conductivity k and geothermal gradient Γ_z in a depth interval of some hundred meters below the surface layer affected by the annual fluctuations of surface temperature. The direct measurement of underground temperature generally requires a temperature-measuring device to be lowered down in a borehole. Logging should be carried out only after the thermal disruption of the bore fluid circulation during drilling is dissipated, i.e. at least 10–20 times the duration of drilling operations. Thermal conductivity analyses of the rock samples collected from each geological formation along the hole are done in laboratory.

Equation (3.10) is based on the assumption that the radiogenic heat does not affect the temperature-depth distribution. Simple calculations show that the error introduced by this assumption is negligible for a depth interval of several hundred meters. In this depth interval, the temperature gradient is normally between 25 and

35 mK m^{-1} and the rock thermal conductivity is 2–3 W m^{-1} K^{-1}. If we extrapolate this gradient to the lithosphere base, a too high temperature should be attained. However, the heat flowing out from the Earth's surface is mainly generated in the crust by the radioactive elements, whose concentration decreases with depth. Consequently, the heat flow and the temperature gradient must decrease with depth.

Geothermal flow measurements assume a perfectly conductive heat flow in a vertical direction. The effects of non-horizontal contacts between formations, whose thermal conductivity laterally varies, distort heat flow from the vertical, and therefore must be removed. Other corrections are sometimes made, to remove the effects of the borehole inclination, topographic relief (ground surface on mountains is generally cooler than that in valleys), climatic changes (long-period fluctuations), and finally sedimentation and uplift (see e.g. Sbrana and Bossolasco 1952; Jessop 1990; Beardsmore and Cull 2001).

The geothermal flow average values in oceanic and continental areas, corrected for the foregoing disturbances, are shown in Table 3.1. Geothermal flow data not representing the regional thermal state, as for example those related to volcanic areas and plate margins, or disturbed by groundwater motion through fracture or permeable formations, are not included. Although the data scatter is very large, undoubtedly the mean geothermal flow decreases with the lithosphere age, as it varies from 30 to 40 mW m^{-2} in shields to >150 mW m^{-2} in the youngest oceanic areas.

The ocean geothermal flow is maximum at the midocean ridges and decreases towards the continents. The decrease is associated with the gradual cooling of magmas after their ascent, the formation of new lithosphere and its progressive growth in thickness. Irregularities may be due to the variation in thermal conductivity and concentration of radioactive elements. The average geothermal flow is about 60 mW m^{-2} in continental areas and 65 mW m^{-2} in oceanic areas (Table 3.2). It is now accepted that the hydrothermal transport due to water

Table 3.1 Geothermal flow versus geological age (after Pasquale 2012)

Oceanic area	Age (10^6 yr)	160–120	120–80	80–40	40–20	20–10	10–0
	mW m^{-2}	50–55	55–60	60–75	75–120	120–150	>150
Continental area	Age (10^6 yr)	>1700	1700–800	800–250	250–50	50–0	
	mW m^{-2}	30–40	40–50	50–60	60–65	65–100	

Table 3.2 Geothermal flow and release of heat

	Geothermal flow mW m^{-2}	Surface 10^6 km^2	Release of heat 10^{12} W
Oceanic area	65 (observed)	310	20 (49 %)
	30 (non-conductive)	310	9 (22 %)
	95 (total)	310	29 (71 %)
Continental area	60 (observed)	200	12 (29 %)

circulating in the uppermost layer of the oceanic crust in areas where the sedimentary cover is thin or even absent is responsible for the dissipation of about 30 mW m^{-2} in a non-conductive manner (Sclater et al. 1980). Therefore, the geothermal flow in oceanic areas should be of 95 mW m^{-2}. On a global scale, the average geothermal flow is then in the order of 80 mW m^{-2}, and this means that in 1 year the heat flowing out from the Earth's interior is equal to about 10^{21} J (or 41 × 10^{12} W). This energy is three orders of magnitude larger than the annual strain energy released from earthquakes as seismic waves.

The main heat sources are the primordial heat originated in the early stages of the Earth's history and radioactivity. The radioactive element concentration strongly decreases with depth and becomes negligible in the Earth's core. In the continental crust the average amount of heat produced by the radioactive decay per unit volume is of about 1 μW m^{-3}. In the oceanic crust, radiogenic heat is less than one-third than that of the continental crust, and in the mantle is on average 0.01 μW m^{-3}. Therefore, the total thermal power due to the radiogenic heat is of the order of 10^{13} W.

The concentration of radioactive elements has also decreased over time and was larger in the Precambrian (Turcotte and Schubert 2002). The radiogenic heat as a function of time can be estimated from the half-life τ (= $\ln 2/\lambda$, where λ is the decay constant) and the present-day concentration of radioactive elements. As a matter of fact, the concentration c_t of an isotope at time t, measured backwards, is related to the present-day concentration c as

$$c_t = c \, \exp(t \ln 2/\tau) \tag{3.11}$$

Thus, the radiogenic heat A versus time (Fig. 3.4) is given by

$$A = \rho \left[\begin{array}{l} 0.993 \, c_U \, A_{U^{238}} \exp(t\ln 2/\tau_{U^{238}}) + 0.0071 \, c_U \, A_{U^{235}} \exp(t\ln 2/\tau_{U^{235}}) \\ + c_{Th} \, A_{Th^{232}} \exp(t\ln 2/\tau_{Th^{232}}) + 0.00012 \, c_K \, A_{K^{40}} \exp(\ln 2/\tau_{K^{40}}) \end{array} \right] \tag{3.12}$$

The heat generation and the half-life of the isotopes are in Table 2.1.
At the present time, heat is mainly produced by uranium and thorium, but 4.5 Gyr ago, when the Earth's crust was forming, potassium could give a contribution comparable to that of uranium because of its shorter half-life.

The original heat is essentially due to the dissipation of gravitational energy and to the short-lived radioactive elements, now extinct, such as Al^{26}. During the subsequent differentiation stage, gravitational energy was again released when droplets of liquid iron or iron-sulfur eutectic trickled down to form the core. It is believed that the core differentiated in a short time, about 0.5 billion years after the formation of the Earth. The latent heat released during the crystallization of the inner core is $T_c \Delta S_c$. With a temperature of the inner core $T_c = 5100$ K, melting entropy $\Delta S_c = 5.8$ J mol^{-1} K^{-1}, molar volume of 4.4×10^{-6} m^3 mol^{-1} and a density of 13000 kg m^{-3} we have a latent heat of about 500 kJ kg^{-1}. The heat flow from the core into the mantle is given by the cooling of the entire core

Fig. 3.4 Radiogenic heat within the Earth as function of time before the present-day. The radiogenic heat at the present-day is set equal to 1: 0.44 from uranium, 0.41 from thorium and 0.15 from potassium

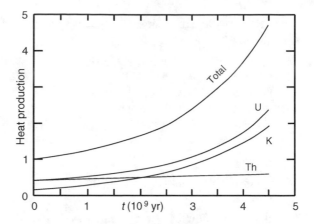

$(2.6 \times 10^{12}$ W), from the crystallization of the inner core $(0.3 \times 10^{12}$ W) and by the gravitational energy $(0.7 \times 10^{12}$ W) (Verhoogen 1980; Poirier 1991). The heat output of the core is therefore about 9 % of the Earth's heat output. Other minor sources are the internal dissipation of energy due to friction caused by tides, which now is negligible, but may have been important in the past when the Moon was closer to Earth, the viscous dissipation due to the mantle convection and the latent heat produced exothermic phase transitions (e.g. olivine-spinel).

It is generally accepted that the present-day geothermal flow of 80 mW m^{-2} is for 60 % generated by the radioactive elements and for 40 % by the cooling of the Earth. The contribution of cooling to the geothermal flow can be estimated from the correspondence between the heat flow output through the surface and the thermal energy decrease inside the Earth. A long-term analysis of the cooling rate of the Earth is given by (Turcotte and Schubert 2002)

$$\frac{dT}{dt} \approx -\frac{3\lambda R T^2}{E^*} \qquad (3.13)$$

where E^* is the activation energy (of the order of 5×10^5 J mol^{-1}), λ is the weighed decay constant of the radioactive elements ($= 2.77 \times 10^{-10}$ yr^{-1}) and R is the universal gas constant ($= 8.314$ J K^{-1} mol^{-1}). For an average temperature of the Earth of 2600 K, (3.13) provides a cooling rate of 93×10^{-9} K yr^{-1}. If q_c is the geothermal flow from cooling, then we have

$$4\pi R_t^2 q_c = -\frac{4}{3}\pi R_t^3 \rho_t c_t \frac{dT}{dt}$$

from which

$$q_c = -\frac{1}{3} R_t \rho_t c_t \frac{dT}{dt} \qquad (3.14)$$

where R_t, ρ_t and c_t are radius, density and specific heat of the Earth, respectively. For $R_t = 6371$ km, $\rho_t = 5515$ kg m^{-3} and $c_t = 900$ J kg^{-1} K^{-1}, we obtain $q_c = 31$ mW m^{-2}. Although estimates by different researchers may be slightly different, there is agreement on the general conclusions presented here. The long-term cooling has important implications for global tectonics. The mantle viscosity increases with time, and consequently convection was more vigorous in the past, when higher temperatures would have yielded a thinner lithosphere.

3.2.2 Geothermal Flow Components

The geothermal flow can be envisaged as a sum of three main components: the heat generated from the radiogenic sources inside the lithosphere, a transient component controlled by tectothermal processes, and the heat which flows out from the asthenosphere. Such a heat flow budget can be made explicit by esti-mating the distribution and amount of the radiogenic heat of the lithosphere and evaluating the thermal perturbation related to tectonic processes.

On a global scale, the heat flow originated by radiogenic heat inside the con-tinental lithosphere, q_r, is about 30 mW m^{-2}. In case of crustal thickening, like over mountain ranges, it increases and can even be twice the normal value. In regions which undergo crustal thinning, q_r is reduced. From heat flow studies of stretched continental margins (Fig. 3.5), Pasquale et al. (1995 and 1996b) argued that q_r is related to crustal thinning according to

$$q_r = 43.2 \, e^{-0.39\beta} \tag{3.15}$$

where the stretching factor β is the ratio between the pre-rift and present-day crustal thickness.

The heat flow component originated by transient perturbations, q_t, yielded by tectonic processes normally is about 30 % of the geothermal flow in Cenozoic regions, whereas it becomes negligible in terrains older than Middle Precambrian (Pollack and Chapman 1977). In extensional realms the heat flow resulting from instantaneous stretching (pure shear mechanism) can be calculated from the relation (McKenzie 1978)

$$q_t = \frac{2 k T_1}{L} \sum_{n=1}^{\infty} \left[\left(\frac{\beta}{n \pi} \right) \sin \left(\frac{n\pi}{\beta} \right) \exp \left(-\frac{n^2 t}{\tau} \right) \right] \tag{3.16}$$

where L is the lithosphere thickness, $\tau = L^2/(\pi^2 \kappa)$ and T_1 is the temperature at the lithosphere base. An empirical expression for the transient component plus the radiogenic heat contribution, q_{rt}, as a function of the relative tinning $(1 - 1/\beta)$ estimated for continental margins is (Pasquale et al. 1996b)

$$q_{rt} = 29.6 + 26.8(1 - 1/\beta) \tag{3.17}$$

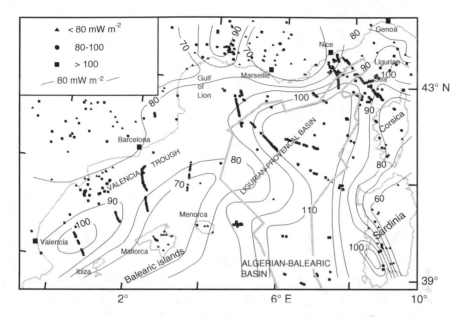

Fig. 3.5 Geothermal flow data distribution and contour map of the NW Mediterranean. The boundary of the oceanic crust is shown by a *grey line* (after Chiozzi 1995; Pasquale et al. 1996b)

The difference between geothermal flow and q_{rt} allows the inference of the contribution of heat flow from the asthenosphere, for any β.

3.2.3 Marginal Basins

Marginal basins are small oceanic basins, underlain by oceanic crust, that represent trapped and isolated fragments of ocean crust or are formed by processes akin to continental rifting or sea-floor spreading (Anderson 1989; Allen and Allen 2005). Although most of the basins are directly related to subduction, the average geothermal flow is on the order of 100 mW m^{-2}. This is high by any standards and difficult to rationalize with the underlying, cold, subducting slab. An overview of the geothermal flow in marginal basins is given by Parson and Sclatter (1977).

Pasquale et al. (1999 and 2005) present a detailed analysis of the geothermal flow of the Tyrrhenian marginal basin (Fig. 1.4). On the continental margins, which stretched 7–11 Myr ago, the average geothermal flow is 125 mW m^{-2} and the stretching factor β ranges between 3.5 and 4.0 (Fig. 3.6). In the portion of the basin underlain by oceanic crust, the geothermal flow does not follow the ocean-floor cooling models, as the average value (160 mW m^{-2}) is lower than that expected. In this case, a purely conductive heat transfer is probably not applicable, and the observed values represent a minimum estimate of the true heat flow.

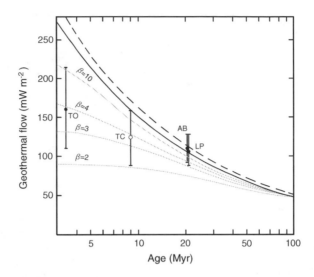

Fig. 3.6 Mean geothermal flow of some Mediterranean basins versus age and ocean-floor cooling models (*black thick curve* by Parsons and Sclater 1977, *black broken curve* by Stein and Stein 1992). The expected geothermal flow on continental lithosphere for different values of the stretching factor β (*grey curves*) is also shown (Pasquale et al. 2005). TO indicates the Tyrrhenian oceanic domain, *TC* the Tyrrhenian continental margin, *AB* and *LP* the oceanic domain of the Algerian–Balearic and the Ligurian–Provençal basins. *Vertical bars* indicate one standard deviation

A convective heat loss of 90–110 mW m^{-2} through hydrothermal circulation within the ocean crust is likely to occur as the theoretically predicted geothermal flow should be 250–270 mW m^{-2} for an oceanic domain with an average age of 3.5 Myr. A possible explanation is that above structural highs, such as seamounts, the absence of thick impermeable sediment might cause cooling by water penetration and circulation.

Differently from the Tyrrhenian basin, in other marginal basin, like e.g. the Algerian–Balearic and the Ligurian–Provençal basins (Figs. 3.5, 3.7), the geothermal flow (105–110 mW m^{-2}) in the oceanic domain matches the ocean-floor cooling models, as the presence of impermeable sediment warrants heat loss to occur mainly by conduction.

The oceanic domains of the young marginal basins are then thermally perturbed, due to strong hydrothermal circulation. This is also a typical feature of oceanic areas at the midocean ridges. Beyond 5-Myr-old oceanic crust, the geothermal flow is less variable and closer to the ocean-floor cooling models. This behavior is associated with an increase in sediment thickness. In fact, when the oceanic crust is covered by a thick blanket of sediment, circulation still occurs, but no heat is lost by convection because the thick sediments are impermeable to sea water. Where the crust is poorly covered, water circulates freely, since the oceanic crust, created by the intrusion of basaltic flows and rapidly cooled dikes, has

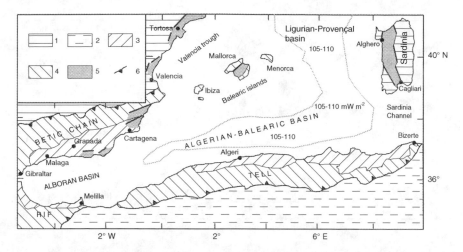

Fig. 3.7 SW Mediterranean basins and main structural elements of the surrounding area. Foreland and foredeep of the European (1) and African (2) plates; internal (3) and external (4) zones of the Betic cordillera and the Rif-Tell chain; Cenozoic rift (5); thrust front (6) (modified after Pasquale et al. 1996c). *Grey line* and *figures* indicate the boundary and geothermal flow range of the oceanic crust

horizontal and vertical cracks on both a large and small scale. In this case, a highly variable geothermal flow results, and the mean heat flow is lower than predicted.

3.2.4 Thrust Sheets

Thrust faults with a very large total displacement are called overthrusts or detachments; these are often found in intensely deformed mountain belts. Geothermal flow can be strongly affected by the overthrusting which occurs at compressive tectonic plate boundaries. Such effects can be estimated with a model that takes into consideration the thickness h of the thrusting slab, the slip rate u and the friction coefficient f at the base of the slab (Brewer 1981; Pasquale et al. 2012).

The geothermal flow at any time t after the start of thrusting is given by the sum of the heat flow arising from thermal relaxation

$$q_{tr} = q_0 \left[1 - \frac{h \exp\left(-\frac{h^2}{4\pi t}\right)}{\sqrt{\pi \kappa t}} \right]$$ (3.18)

and from frictional heating

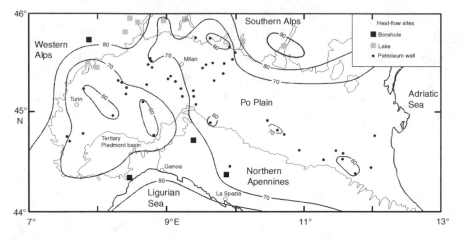

Fig. 3.8 Contour map of geothermal flow (isolines in mW m^{-2}) of N Italy

$$q_f = k \frac{u\,\tau}{\rho\,c_p\,\kappa}\,\mathrm{erfc}\left(\frac{h}{2\sqrt{\kappa t}}\right) \qquad\qquad\qquad t \leq t_1$$

$$q_f = k \frac{u\,\tau}{\rho\,c_p\,\kappa}\left[\mathrm{erfc}\left(\frac{h}{2\sqrt{\kappa t}}\right) - \mathrm{erfc}\left(\frac{h}{2\sqrt{\kappa(t-t_1)}}\right)\right] \qquad t > t_1$$

$$(3.19)$$

where ρ, k, κ and c_p are density, thermal conductivity, thermal diffusivity and specific heat of the rock, respectively, and q_o is the geothermal flow before thrusting. The rate of heating is proportional to the stress across the fault: $\tau = f\rho g h$, where g is acceleration due to gravity and t_1 is the time of thrusting.

Pasquale et al. (2010 and 2012) investigated the thermal effects of overthrusting in the Po Plain (Italy) and demonstrated that the geothermal flow is sensibly reduced because of the recent thrust (Fig. 3.8). The geothermal flow variation with time, $q_o(t) = q_{tr} + q_f$, obtained from (3.18) and (3.19) for a slab that slips

Fig. 3.9 Geothermal-flow variation arising from overthrusting at different rates of slip of the Apennines buried unit. Model parameters: $g = 9.8$ m s^{-2}, $\kappa = 30$ m^2 yr^{-1}, $k = 2.5$ W m^{-1}K^{-1}, $c_p = 1000$ J kg^{-1}K^{-1}, $\rho = 2700$ kg m^{-3} and $f = 0.6$

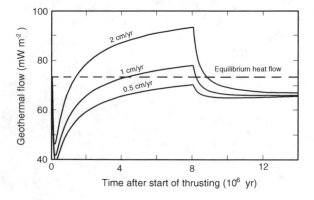

horizontally at different constant rates is depicted in Fig. 3.9. Geometry and timing of the deformation of the thrusts are referred to the youngest compressional tectonic event that has taken place from the Messinian to the Pleistocene. For a slab with a thickness of 5 km and a geothermal flow equilibrium value of 73 mW m^{-2}, the heat flow initially decreases by about 35–45 %. Then, it increases until the end of the thrusting, reaching values of about 65–70 mW m^{-2} for a slip rate of 0.5 cm yr^{-1} and 90–95 mW m^{-2} for a slip rate of 2 cm yr^{-1}. Subsequently, the heat flow decreases only due to thermal relaxation. The geothermal flow observed is better matched for a slip rate of 1 cm yr^{-1}.

3.3 Temperature Distribution in the Lithosphere

The lithosphere thermal regime is dominated by conduction except in areas affected by magmatic and hydrothermal processes. The temperature distribution can be determined by solving the general equation of heat conduction, considering the fact that the thermal conditions differ from continental to oceanic plates. In the former, there is often no variation of temperature with time, at least in shields and cratons, and the radiogenic heat is high. In the latter, there is significant advective cooling of the lithosphere, which moves away from the ridge, whereas radiogenic heat is negligible.

3.3.1 Continental Plate

By assuming that temperature is a function only of depth z (positive downwards), the heat conduction equation (2.21) in a one-dimensional form becomes

$$k\,\frac{d^2T}{dz^2} + A_o \exp\left(-\frac{z}{D}\right) = 0 \tag{3.20}$$

which integrated gives

$$k\,\frac{dT}{dz} - D\,A_o \exp\left(-\frac{z}{D}\right) = c_1 \tag{3.21}$$

where the thermal conductivity k is independent of temperature and the surface radiogenic heat A_o decreases exponentially with depth according to (2.60). Since at the base of the plate the second term of the first member is negligible, the integration constant c_1 can be taken as equal to the heat flow from the asthenosphere $q_a\,(= -q)$. Equation (3.21) then becomes

$$q = -q_a - D\,A_o \exp\left(-\frac{z}{D}\right) \tag{3.22}$$

which gives the heat flow at any depth. The quantity q_a is also called reduced heat flow. Since $q = -q_o$ at $z = 0$, we find the linear relationship (2.61) between geothermal flow and surface radiogenic heat. Equation (3.22) holds in most of the continental areas, with q_a varying from 20 mW m^{-2} (Precambrian areas) to 45 mW m^{-2} (Cenozoic areas).

By integrating (3.21), one has

$$kT + D^2 A_o \exp\left(-\frac{z}{D}\right) - q_a z = c_2 \tag{3.23}$$

where the constant of integration c_2 can be determined from the boundary condition $T = T_o$ at $z = 0$, namely $c_2 = kT_o + D^2 A_o$. We can then write

$$T = T_o + \frac{D^2 A_o}{k}\left[1 - \exp\left(-\frac{z}{D}\right)\right] + \frac{q_a}{k} z \tag{3.24}$$

This is the solution of the heat conduction equation for a continental plate, which allows the calculation of $T(z)$. Figure 3.10 shows some geotherms for different values of geothermal flow. The temperature at the Moho varies from a minimum of 350 °C in Precambrian shields to over 800 °C beneath some Cenozoic orogens. The intersection between geotherms and the curve $0.85 T_s$, where T_s is the solidus of the upper mantle, provides a good estimate of the thickness of the lithosphere (Pollack and Chapman 1977). Beneath the continents, the thickness of the lithosphere ranges from about 50 km in young areas, with large geothermal flow, to 250–350 km in Precambrian shields where there is a total absence of a low-velocity zone. This conclusion strengthens the results obtained from the studies of seismic surface waves dispersion. On the basis of the geothermal flow and a petrological model of the crust and upper mantle (see Sect. 2.6), the temperatures within a plate can be calculated with an accuracy of 100 °C. Better estimates are

Fig. 3.10 Continental geotherms for different values of geothermal flow. The intersection of the curve $0.85 T_s$ with the geotherms corresponds to lithospheric thickness. $k = 3.0$ W K^{-1}m^{-1}, $D = 10$ km and $A_o = 2.5$ µW m^{-3}

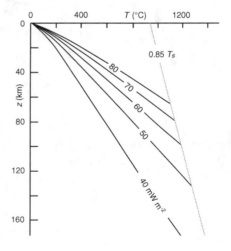

obtained if the pressure and, especially, temperature dependence of thermal conductivity are taken into account (see e.g. Pasquale 1987; Pasquale et al. 1990).

3.3.2 Oceanic Plate

Thermal conditions of an oceanic plate are different from those of a continental plate, since the material originated at the ridges cools as it moves away and the radiogenic heat in the crust is negligible. If x and z are the horizontal and vertical coordinates, respectively, and $v = dx/dt$ is the plate velocity, assumed to be constant and in practice determined by the age t of the crust at a distance x from the ridge, we get

$$\frac{\partial T}{\partial t} = \frac{\partial T}{\partial x}\frac{dx}{dt} = v\frac{\partial T}{\partial x} \tag{3.25}$$

Furthermore, since horizontal heat conduction is negligible if compared to the transport of heat caused by the plate horizontal motion, we have

$$\nabla^2 T = \frac{\partial^2 T}{\partial x^2} + \frac{\partial^2 T}{\partial z^2} \approx \frac{\partial^2 T}{\partial z^2} \tag{3.26}$$

hence the equation of heat conduction under transient regime (2.23) becomes

$$\frac{\partial^2 T}{\partial z^2} = \frac{v}{\kappa}\frac{\partial T}{\partial x} \tag{3.27}$$

The problem of finding $T(x, z)$ in a moving plate is then equivalent to finding $T(z, t)$ in a stationary plate. If one considers that plate cooling is similar to a semi-infinite solid initially at temperature T_s and with the surface maintained at temperature T_o, the solution of (3.27) is

$$\frac{T - T_o}{T_s - T_o} = \mathrm{erf}\left(\frac{z}{2\sqrt{\kappa\,t}}\right) = \mathrm{erf}\left(\frac{z}{2\sqrt{\kappa\,x/v}}\right) \tag{3.28}$$

The oceanic lithosphere can be considered as a moving thermal boundary layer in which cooling takes place by conduction and where the temperature gradient is higher than the convecting substratum. The lithosphere thickness as a function of age (or distance from the ridge) can be obtained from (3.28). By assuming that the temperature at the lithosphere base is $0.85T_s$ ($T_s = 1550\,°C$ for a dry peridotite at upper mantle pressure) and $T_o = 5\,°C$, it is possible to obtain

$$\frac{0.85\,T_s - T_o}{T_s - T_o} = \mathrm{erf}\left(\frac{h_t}{2\sqrt{\kappa\,t}}\right) = 0.849 \tag{3.29}$$

where h_t is the thickness of the lithosphere. From (3.29) it follows that $h_t/(4\chi t)^{1/2} = 1.02$. Thus

$$h_t = 2.0 \sqrt{\kappa\, t} \tag{3.30}$$

or

$$h_t = 2.0 \sqrt{\kappa\, x/v} \tag{3.31}$$

For $t \leq 80 - 100 \times 10^6$ yr, (3.30) agrees with the evidence provided by the study of seismic surface wave dispersion. An older lithosphere tends to have a constant thickness, which is probably evidence of a heat source in the upper part of the asthenosphere related to thermal convection and/or shear heating. Isotherms, geotherms and the h_t curve determined from (3.28) and (3.30) are shown in Fig. 3.11.

The relationship between h_t and t (or x) is only one of several laws that may be applied to the oceanic lithosphere. Another relation is that between geothermal flow and lithosphere age. According to (3.9), (3.28) can be rewritten as

$$T = T_0 + (T_s - T_0)\frac{2}{\sqrt{\pi}} \int_0^{z/\sqrt{4\kappa t}} e^{-u'^2}\, \mathrm{d}u' \tag{3.32}$$

which, by differentiating with Leibniz rule, provides the thermal gradient

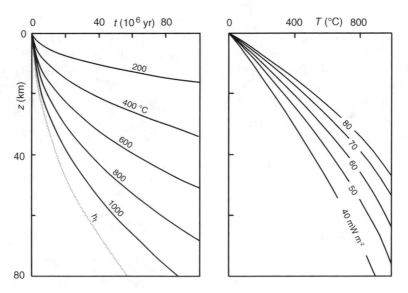

Fig. 3.11 Isotherms and geotherms in the oceanic lithosphere. $\kappa = 30$ m^2 yr^{-1}, v $=$ 3 cm yr^{-1}, $k = 3$ W K^{-1}m^{-1}. Lithosphere thickness h_t from (3.30)

Fig. 3.12 Geothermal flow
pattern perpendicular to the
ridge. The *grey curve* is given
by (3.34). Geothermal flow
data after Stein and Stein
(1992)

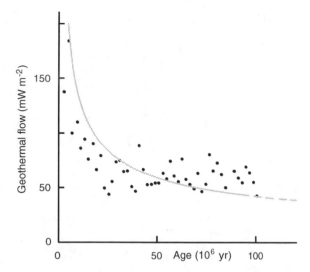

$$\frac{\mathrm{d}T}{\mathrm{d}z} = \frac{T_s - T_\mathrm{o}}{\sqrt{\pi \, \kappa \, t}} \exp\left(-\frac{z^2}{4\kappa \, t}\right) \tag{3.33}$$

By introducing this gradient for $z = 0$ in (3.10), we find

$$q_\mathrm{o} = k \frac{T_s - T_\mathrm{o}}{\sqrt{\pi \, \kappa \, t}} = k \, (T_s - T_\mathrm{o}) \left(\frac{\mathrm{v}}{\pi \, \kappa \, x}\right)^{1/2} \tag{3.34}$$

The decrease of geothermal flow q_o with age (or with the distance from the ridge
axis) expressed by (3.34) is confirmed by data (Fig. 3.12). However, for $t \leq 40$
Myr the observed q_o is less than expected. This is due to convective cooling caused
by hydrothermal circulation in the crust next to the oceanic ridges (see
Sect. 3.2.3). When age increases, the sedimentary cover (less permeable than
fractured basalts) prevents the seawater infiltration into the oceanic crust and
consequently convective cooling. For $t \geq 80$ Myr, the observed q_o reaches an
asymptotic value, given by the heat flow from the asthenosphere plus the heat flow
produced by the radioactive elements of the oceanic plate.

3.3.3 Ocean Floor Topography

Equation (3.28) can also be used to predict the ocean floor topography. Since the
oceanic lithosphere is approximately in isostatic equilibrium, the average seafloor
depth should increase with age as a consequence of the density increase caused by
cooling. The principle of isostasy states that there is the same mass per unit area
between the surface and the depth of compensation for any vertical column of

Fig. 3.13 Ocean lithosphere
under the ridge

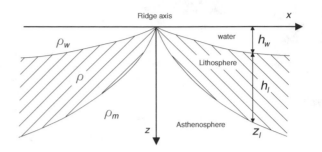

material (Fig. 3.13). Thus, by indicating with h_w the depth of the ocean floor, h_l and ρ the thickness and density of the lithosphere, ρ_w and ρ_m the densities of water and asthenospheric mantle, respectively, we obtain

$$\rho_m \, (h_w + h_l) = \rho_w \, h_{\mathbf{w}} + \int_0^{h_l} \rho \, \mathrm{d}z \qquad (3.35)$$

or

$$h_w \, (\rho_m - \rho_w) = \int_0^{h_l} (\rho - \rho_m) \, \mathrm{d}z \qquad (3.36)$$

where $\rho > \rho_m$ because of the thermal contraction of the cooling lithosphere.

According to the definition of the expansion coefficient α, we can write

$$\rho - \rho_m = \alpha \, \rho_m \, (T_s - T) \qquad (3.37)$$

so that (3.36) becomes

$$h_w \, (\rho_m - \rho_w) = \alpha \, \rho_m \int_0^{h_l} (T_s - T) \, \mathrm{d}z \qquad (3.38)$$

By substituting in this equation the temperature T from (3.28) we obtain

$$h_w \, (\rho_m - \rho_w) = \alpha \, \rho_m \, (T_s - T_o) \int_0^{\infty} \left[\mathrm{erfc} \left(\frac{z}{2 \, \sqrt{\kappa \, t}} \right) \right] \mathrm{d}z \qquad (3.39)$$

where the limit on the integral has been changed from $z = h_l$ to $z = \infty$, since $\rho \to \rho_m$ and $T \to T_s$ at the base of the lithosphere. By using the variable $u = z/(4\kappa t)^{1/2}$, (3.39) becomes

$$h_w(\rho_m - \rho_w) = 2\alpha\rho_m(T_s - T_o)(\kappa t)^{1/2} \int_0^{\infty} \mathrm{erfc}(u)\mathrm{d}u' \qquad (3.40)$$

Moreover, as

$$\int_0^\infty \mathrm{erfc}(u)\mathrm{d}u' = \frac{1}{\sqrt{\pi}}$$

the ocean floor depth h_w as a function of age t or distance x from the ridge is given by

$$h_w = \frac{2\alpha\rho_m(T_s - T_o)}{\rho_m - \rho_w}\left(\frac{\kappa t}{\pi}\right)^{1/2} = \frac{2\alpha\rho_m(T_s - T_o)}{\rho_m - \rho_w}\left(\frac{\kappa x}{\pi v}\right)^{1/2} \tag{3.41}$$

This equation states that if isostatic equilibrium conditions prevail, the average depth of the ocean will increase with the square root of the age or of the distance from the ridge. This conclusion is especially verified by observations for $t \le 80 - 100 \times 10^6$ yr.

In conclusion, the model of oceanic lithosphere as a thermal boundary layer that moves away from the ridge allows us to estimate the temperature and thickness of the lithosphere with (3.28) and (3.30), whereas the geothermal flow and the depth of the seafloor can be inferred from (3.34) and (3.41), respectively.

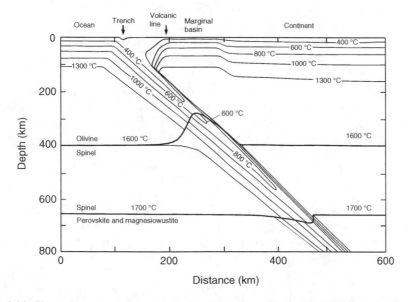

Fig. 3.14 Thermal structure of the descending lithosphere. The lower temperature in the slab causes the elevation of the olivine-spinel phase boundary (modified from Schubert et al. 2001)

3.4 Thermal Regime of the Deeper Interior

Estimates of the temperature at the lithosphere base range from 1250 to 1350 °C. Along the spreading ridges and in volcanic areas, isotherms rise while in areas with subducted lithosphere isotherms are distorted downwards (Fig. 3.14). Below the lithosphere, a seismic low-velocity zone, 50–150 km thick, occurs. In this zone, the presence of small amounts (0.1–1 %) of liquid basalt may justify the decrease by 5 % of seismic wave velocity, and in particular for shear waves. The occurrence of liquid basalt can be accounted for by a lowering of the mantle solidus due to the presence of water and volatiles, which are released through dehydration reactions of amphiboles.

The mantle is not only heated by radioactive elements. Since the pressure at great depths is enormous, there must be an increase in temperature resulting from compression. The volume reduction of the material due to pressure produces heat and, consequently, an increase in temperature, since the process does not exchange heat with the surroundings (adiabatic compression). If the material were incompressible, there would be no increase in temperature. However, mantle is sufficiently compressible to originate adiabatic temperature gradients. According to recent results on the specific heat and expansion coefficient at high pressure and temperature, a temperature of 1200 °C only due to adiabatic compression would be expected at the Gutenberg discontinuity. Studies on the relationship between pressure and temperature on the olivine—spinel and spinel—post-spinel transitions provided the benchmark temperatures of 1600 °C at 400 km and 1700 °C at 670 km, with uncertainties of the order of 100 °C.

The melting temperature T_m fixes the possible maximum temperature at the mantle base. However, the estimate of T_m is subject to large uncertainty, since it is experimentally unknown at high pressures (although experiments with short shock waves can reach the pressures of the core). Among the different laws proposed for T_m as a function of pressure, the most reliable is that of Lindemann (see Poirier 1991). He considered a solid formed by simple harmonic oscillators, distributed in a cubic crystal lattice, and demonstrated that solid melts when the thermal oscillation of atoms reaches a critical amplitude

$$T_m = C \ m \ V^{2/3} \vartheta_D^2 \qquad (3.42)$$

where C is a constant, m the mass of the atoms, V the volume and ϑ_D the Debye temperature. Other theoretical and experimental equations have been also proposed; many of them are modifications of (3.42). For the calculation of T_m at pressure p_m, we can use the equation (Gilvarry 1956)

$$\frac{p_m}{p_o} = \left(\frac{T_m}{T_{m_o}}\right)^c - 1 \qquad (3.43)$$

where T_{m_o} as eq. 3.43 is the melting temperature at zero pressure, p_o is the (negative) melting pressure at zero temperature and $c = (6\gamma + 1)/(6\gamma - 2)$ where

γ is the Grüneisen parameter. From similar considerations, Stacey (1992) estimated that T_m at the mantle base is $3500\,^{\circ}\mathrm{C}$. The core essentially consists of iron, which is molten in the outer core and solid in the inner core. At the inner core boundary, the melting temperature of iron obtained theoretically or through the extrapolation of experimental data varies over a large interval, although generally ranging around $5000\,^{\circ}\mathrm{C}$. Table 3.3 gives the values of physical properties and the temperature T of the principal discontinuities of the Earth's interior, as estimated by Stacey (1992). The closeness of T to the melting temperature, T_m, is an indication of material viscosity. The Debye temperature, T_D, increases with pressure, being determined by the frequencies of the modes of crystal lattice vibration, which are related to the elastic moduli.

The lithosphere drift proves that mantle can convect. Also, the thermal conductivity is too low to remove heat only through conduction. If the Earth were able to dissipate heat only through conduction, it would cool according to

$$T(z,t) = T_1 \mathrm{erf}(u) \tag{3.44}$$

where T_1 is initial temperature and the variable $u = z(4\kappa t)^{-1/2}$ describes the relationship between the distance and the characteristic thermal time τ of propagation of heat by conduction. Since $\mathrm{erf}(u) \approx 0.5$ for $u = 0.5$, the temperature at depth z would be equal to $0.5T_1$ for a time $\tau = z^2/\kappa$. For example, at a depth of 500 km and $\kappa = 30$ m^2 yr^{-1}, the value of τ is much longer than the Earth's current age, i.e. is of the order of 8 Gyr. Consequently, heat at depth can not be removed only through conduction. It follows that heat must be transported by thermal convection, in other words by a mechanism involving mass transfer.

In a convective system, the temperature gradient is slightly higher than the adiabatic gradient except at thermal boundary layers. The adiabatic lapse rate is defined as the rate of increase in temperature with depth resulting from an increase in pressure. The adiabatic temperature gradient can be calculated from the thermodynamic relation between entropy per unit mass, S, temperature and pressure: $\mathrm{d}S = c_p \mathrm{d}T/T - \alpha \mathrm{d}p/\rho$ from which, for an adiabatic, reversible process ($\mathrm{d}S = 0$), it follows that

$$\left(\frac{\mathrm{d}T}{\mathrm{d}p}\right)_S = \frac{\alpha\, T}{c_p\, \rho} = \frac{\gamma\, T}{K_S} \tag{3.45}$$

where γ is the Grüneisen parameter, c_p is the specific heat, α is the expansion coefficient and K_S is the adiabatic bulk modulus. Since, for a homogeneous mantle, $\mathrm{d}p/\mathrm{d}z = \rho g$, the increase of the adiabatic temperature with depth takes the form of

$$\left(\frac{\mathrm{d}T}{\mathrm{d}z}\right)_S = \left(\frac{\mathrm{d}T}{\mathrm{d}p}\right)_S \frac{\mathrm{d}p}{\mathrm{d}z} = \frac{\alpha\, g\, T}{c_p} = \frac{\gamma\, \rho\, g\, T}{K_S} = \frac{\gamma\, g\, T}{\phi} \tag{3.46}$$

where ϕ is the seismic parameter. The adiabatic temperature gradient in the upper mantle is of the order of 0.4–0.5 mK m^{-1} and at greater depths decreases (0.25–0.30 mK m^{-1}). The onset of convection in the mantle implies that the

Table 3.3 Density ρ, pressure p, seismic parameter ϕ, electrical conductivity σ_e, thermal conductivity k, Grüneisen parameter γ; specific heat c_p, expansion coefficient α, Debye temperature T_D, melting temperature T_m and temperature T in correspondence of the major discontinuities

z km	ρ kg m⁻³	p GPa	ϕ km² s⁻²	σ_e S m⁻¹	k W m⁻¹ K⁻¹	γ	c_p J kg⁻¹ K⁻¹	α 10⁻⁶ K⁻¹	T_D	T_m	T K
Upper mantle											
100	3370	3	38	0.01	5	0.5	1200	16	700	1800	1500
400	3540	13	49	0.01	5	0.6	1250	16	850	2100	1850
Transition zone											
400	3720	13	51	0.05	5	0.6	1250	15	850	2100	1850
670	3990	24	64	0.1	5	0.7	1250	15	900	2750	1950
Lower mantle											
670	4380	24	69	1.0	7	1.0	1275	19	1000	2750	1950
2741	5490	127	117	30	10	1.0	1275	9	1450	3800	3000
Layer D''[a]											
2741	5490	127	117	30	10	0.8	1250	9	1450	3850	3550
2891	5560	136	65	50	10	0.8	1250	9	1400	3850	3750
Outer core											
2891	9900	136	65	3 × 10⁵	28	1.4	700	16	–	3450	3750
5150	12170	329	107	3 × 10⁵	37	1.3	650	8	–	4950	4950
Inner core											
5150	12760	329	105	4 × 10⁵	50	1.1	650	7	1300	4950	4950
6371	13090	364	109	4 × 10⁵	50	1.1	650	6	1350	5200	5100

[a] D'' is a thermal boundary layer in which heat is transferred only by conduction, thus requiring a steeper gradient than in the convecting regions situated above and below

temperature gradient must be superadiabatic. Temperatures given in Table 3.3 indicate that the gradient is decidedly superadiabatic.

The definition of adiabatic gradient provides an estimate of the variation of the adiabatic temperature. The integration of (3.45) gives

$$\ln\left(\frac{T_2}{T_1}\right) = \gamma \int_{p_1}^{p_2} \frac{dp}{K_S} \tag{3.47}$$

where γ is assumed to be approximately constant. By using the definition of $K_S(= -V dp/dV = \rho dp/d\rho)$, it is possible to have

$$\frac{T_2}{T_1} = \left(\frac{\rho_2}{\rho_1}\right)^{\gamma} \tag{3.48}$$

which gives the variation of adiabatic temperature through a finite interval of depth in terms of density ratio.

A necessary, but not sufficient, condition for convective instability is that the temperature gradient in the mantle $(\vartheta_2 - \vartheta_1)/h$ (upper boundary, $z = 0$, at temperature ϑ_1, and the lower boundary, $z = h$, at temperature ϑ_2) is larger than the adiabatic gradient. A tighter criterion is given by the dimensionless Rayleigh number, defined as

$$R_a = \frac{g \, \alpha \, \rho \, h^3 (\vartheta_2 - \vartheta_1)}{\eta \, \kappa} \tag{3.49}$$

where $\vartheta_2 > \vartheta_1$, and the dynamic viscosity η, for a mantle assumed to behave as a Newtonian viscous fluid, is given by the ratio between shear stress and strain rate. For $\eta = 10^{21}$ Pa s, $g = 10$ m s^{-2}, $\alpha = 12 \times 10^{-6}$ K^{-1}, $\rho = 4200$ kg m^{-3}, $\kappa = 10^{-6}$ m^2 s^{-1} and $(\vartheta_2 - \vartheta_1)/h = 2000/2600$ K km^{-1}, the Rayleigh number equals 2×10^7, i.e. it is considerably larger than the critical value (about 2000). The strain rate or velocity gradient in the entire mantle convection is of the order of 10^{-15} s^{-1}, for a convective cell radius of 1300 km and a velocity of 10^{-9} m s^{-1} (3 cm yr^{-1}). For a convective mechanical power of 10^{12} W, which deforms the mantle and corresponds to 10 % of the available power, and a volume of 10^{21} m^3, the stress needed to maintain this constant strain rate is 1 MPa.

The classic models assume convection to occur in the whole mantle, but many observations (thermodynamical, geochemical) can also be explained in terms of convection within two separate layers. It is likely that the upper and lower mantle are two different chemical reservoirs that separately form the continental crust and the core (Allègre 1982). The Rayleigh number is supercritical in both reservoirs. Some numerical and laboratory experiments brought to the conclusion that two-layer convection is possible if the density contrast due to changes in the chemical composition is larger than the changes caused by temperature. A separate convection requires a thermal boundary layer placed between the upper and lower mantle, through which heat is transferred by conduction (Fig. 3.15). For a wider

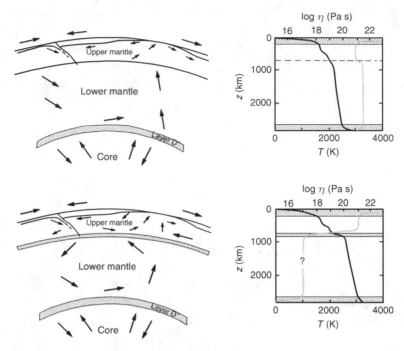

Fig. 3.15 Temperature (*black curve*) and viscosity (*grey curve*) profiles (data from Poirier 1991) for mantle-wide and two-layer convection models. *Dotted*, superadiabatic thermal boundary layers. Deviations from a spherical symmetry do not occur only in the lithosphere, but affect the entire mantle, since the differences in temperature between hot material (that rises) and cold material (that drops) in a convective system trigger horizontal temperature gradients

discussion of the transferred heat and convection patterns in the mantle, the reader can refer to Poirier (1991) and Schubert et al. (2001).

Factors that could inhibit a mantle-wide convection are a viscosity increase with depth, a phase change with an inclination of the Clapeyron curve sufficiently negative and the lack of mixing between chemically different layers. Geophysical evidence indicates an increase in viscosity. This is also confirmed by micro-physical evidence. The slope of the Clapeyron curve (the equilibrium curves p–T separating two phases of the same material) is positive $(dp/dT \approx 4 \text{ MPa K}^{-1})$ for the transition olivine-spinel and therefore the phase interface at depth of 400 km is shifted upward in the descending currents and down in the ascending ones. These density perturbations would help circulation, and therefore phase changes are not an obstacle to convection.

Further information can be deduced from the distribution of the electrical conductivity σ_e (Stacey 1992). The conductivity of the surface rocks is mainly electrolytic (motion of ions) and is therefore linked to pore fluids according to Archie's law

$$\sigma_e = \sigma_{ef}\, \xi^n \tag{3.50}$$

where σ_{ef} is the conductivity of the fluid, ξ the fraction of pores filled with the fluid and n a parameter that increases with compaction, cementation and consolidation and varies between 1.3 and 2.5. Sands have values between 1.3 and 1.5 (Schön 1996). Generally, the unconsolidated sediments have a conductivity of 0.1–5.0 S m^{-1}, while the upper crust rocks, characterized by low porosity, are less conductive (0.001–0.01 S m^{-1}); temperature favors the circulation of fluids and increases electrical conductivity, whereas pressure, which reduces voids, tends to decrease it.

In the lower crust and at greater depths, ionic conduction takes place, i.e. ions move through defects in the mineral crystal lattice, together with conduction through impurities and electronic conduction (ohmic). In this case, the relation between temperature and σ_e is described by

$$\sigma_e = \sigma_{e\infty}\exp\left(-\frac{E^*}{k_B\,T}\right) \tag{3.51}$$

where E^* is the activation energy, k_B the Boltzmann constant, T the absolute temperature and $\sigma_{e\infty}$ the extrapolated conductivity for $T \rightarrow \infty$. Electric conductivity values in the mantle and the core are listed in Table 3.3. Various factors, besides temperature, affect uncertainties in conductivity, such as partial melting, the presence of water and volatile impurities.

Since it is now accepted that the main magnetic field is generated by the electric currents, which circulate in the outer core and are driven by a magneto-hydrodynamic dynamo, a large value of conductivity (about 10^5 S m^{-1} with a probable error of one order of magnitude) can be assumed for the core. Since both the thermal and electrical conduction in the inner core are dominated by the electron contributions, the following simple relationship (the Wiedemann–Franz law) is valid

$$\frac{k}{\sigma_e} = L\,T \tag{3.52}$$

where T is absolute temperature and $L = 2.5 \times 10^{-8}$ W S^{-1}K^{-2} is the Lorentz number (Stacey 1992). With values $k = 50$ W m^{-1}K^{-1} and $T = 5000$ K, we find $\sigma_e = 4 \times 10^5$ S m^{-1}.

References

Allègre CJ (1982) Chemical geodynamics. Tectonophysics 81:109–132
Allen PA, Allen JR (eds) (2005) Basin analysis: principle and applications, 2nd edn. Blackwell, Oxford
Anderson DL (1989) Theory of the Earth. Blackwell, Oxford
Beardsmore GR, Cull JP (2001) Crustal heat flow: a guide to measurement and modelling. Cambridge University Press, Cambridge

Brewer J (1981) Thermal effects of thrust faulting. Earth Planet Sci Lett 56:233–244
Carslaw HS, Jaeger JC (1986) Conduction of heat in solids, 2nd edn. Clarendon Press, Oxford
Chiozzi P (1995) Subsidenza, regime termico e domini crostali nel Mediterraneo nordoccidentale.
    Ph D Thesis, Genoa University, Italy
Gilvarry JJ (1956) Grüneisen's law and the fusion curve at high pressure. Phys Res 102:317–325
Jessop AM (1990) Thermal geophysics. Elsevier, Amsterdam
McKenzie D (1978) Some remarks on the development of sedimentary basins. Earth Planet Sci
    Lett 40:25–32
Parsons B, Sclater JG (1977) An analysis of the variation of ocean floor bathymetry and heat low
    with age. J Geophys Res 82:803–827
Pasquale V (1987) Possible thermal structure of the eastern part of the Central Alps. Nuovo
    Cimento 10C:129–141
Pasquale V (2012) Geofisica. ECIG-Edizioni Culturali Internazionali Genova, Genova
Pasquale V, Cabella C, Verdoya M (1990) Deep temperatures and lithospheric thickness along
    the European geotraverse. Tectonophysics 176:1–11
Pasquale V, Verdoya M, Chiozzi P (1995) On the heat flux related to stretching in the NW-
    Mediterranean continental margins. Studia Geoph Geod 39:389–404
Pasquale V, Verdoya M, Chiozzi P (1996a) Climatic signal from underground temperatures.
    International conference on Alpine meteorology, Hydrometeorological Institute of Slovenia,
    Bled, pp 201–208
Pasquale V, Verdoya M, Chiozzi P (1996b) Heat flux and timing of the drifting stage in the
    Ligurian-Provençal basin (NW Mediterranean). J Geodyn 21:205–222
Pasquale V, Verdoya M, Chiozzi P (1996c) Some geophysical constraints to dynamic processes
    in the Southwestern Mediterranean. Ann Geofis 39:1185–1200
Pasquale V, Verdoya M, Chiozzi P (1998) Climate change from meteorological observations and
    underground temperature in Northern Italy. Studia Geoph Geod 42:30–40
Pasquale V, Verdoya M, Chiozzi P (1999) Thermal state and deep earthquakes in the Southern
    Tyrrhenian. Tectonophysics 306:435–448
Pasquale V, Verdoya M, Chiozzi P (2005) Thermal structure of the ionian slab. Pure Appl
    Geophys 162:967–986
Pasquale V, Chiozzi P, Verdoya M (2010) Tectonothermal processes and mechanical strength in
    a recent orogenic belt: Northern Apennines. J Geophys Res 115:148–227
Pasquale V, Chiozzi P, Verdoya M, Gola G (2012) Heat flow in the Western Po Basin and the
    surrounding orogenic belts. Geophys J Int 190:8–22
Poirier JP (1991) Introduction to the physics of the Earth's interior. Cambridge University Press,
    Cambridge
Pollack HV, Chapman DS (1977) On the regional variation of heat flow, geotherms and
    lithospheric thickness. Tectonophysics 38:279–296
Sbrana F, Bossolasco M (1952) Sul regime termico degli strati superiori della crosta terrestre.
    Geofis Pura Appl 23:21–26
Sclater JG, Jaupart C, Galson D (1980) The heat flow through the oceanic and continental crust
    and the heat loss of the Earth. Res Geophys Space Phys 18:269–311
Schön J (1996) Physical properties of rocks: fundamentals and principles of petrophysics.
    Handbook of geophysical exploration, vol 18. Redwood Books, Trowbridge
Schubert G, Turcotte DL, Olson P (2001) Mantle convection in the Earth and planets. Cambridge
    University Press, Cambridge
Stacey FD (1992) Physics of the Earth, 3rd edn. Brookfield Press, Brisbane
Stein C, Stein S (1992) A model for the global variation in oceanic depth and heat flow with
    lithospheric age. Nature 359:123–128
Turcotte D, Schubert GL (2002) Geodynamics—application of continuum physics to geological
    problems, 2nd edn. Cambridge University Press, Cambridge
Verhoogen J (1980) Energetics of the Earth. National Academy of Sciences, Washington DC

Chapter 4
Temperature and Magmatic Processes

Abstract Physical or chemical instabilities in the Earth's interior may cause the formation of magma, phenomenon which consequently creates a mechanical disequilibrium with the surrounding rocks. The magmatic fluid attempts to find an equilibrium condition through energy and mass transfer from the bottom upwards. The volcanic activity is therefore the surface expression of a redistribution of the internal energy of the Earth through an advection mechanism. This chapter focuses on the formation processes and upwelling mechanisms of magma and on the parameters that control its rheological behavior. Solidification, involving phase change, and cooling of intrusive igneous bodies and lava covers are also dealt with.

Keywords Magma genesis · Magmatic underplating · Magma rheology · Upwelling mechanisms · Solidification processes · Cooling models

4.1 Melting Mechanisms

Present-day physical models of the melting processes that occur in the mantle, which produce the lava erupted by volcanoes and the igneous rocks, have reached high levels of sophistication, but some intriguing questions still emerge. One particular puzzler is the occurrence of magmatism where there is little tectonic extension and where there is no obvious relationship to an anomalously hot plume of mantle like those beneath oceanic hotspots. This section presents the most reliable hypothesis on the formation of magma, and in particular focuses on the emplacement of magmatic fluids at the base of the continental crust due to stretching processes.

V. Pasquale et al., *Geothermics*, SpringerBriefs in Earth Sciences,
DOI: 10.1007/978-3-319-02511-7_4, © The Author(s) 2014

4.1.1 Solidus Intersection

Since the formation of magma cannot be directly observed, it must be based on the knowledge of the phenomena observable at the Earth's surface and on laboratory studies of chemical and physical processes, whose results are extrapolated to high pressure and temperature conditions. The fundamental problem of the study of magma formation is to explain why a portion of the mantle can melt and form a molten rock or rather a silicate liquid (magma) containing suspended solids (such as crystals and rock fragments) and gas phases.

In the portions of the Earth's interior that are subject to convection, the geotherms tend to the same adiabatic curve. This means that under normal conditions the mantle is still solid and its melting is an exceptional event (Fig. 4.1). The condition that may cause mantle melting is the intersection of the geotherms with the mantle solidus. This may occur through several mechanisms: pressure decrease, temperature rise, depression of mantle solidus and variations of pressure and temperature caused by metasomatic processes.

A decrease in pressure could be in principle caused by a fracture in the crust that reaches the mantle. However, the pressure at depths larger than 12 km is so high that a fracture cannot remain open (Zoback and Townerd 2001). Another possibility is a pressure decrease due to the rise towards the surface of a portion of mantle. Due to the low thermal conductivity of the mantle, convection drags material nearly adiabatically. In other words, a mantle portion rises to a shallower depth so that it can generate magma within the ascending branch of convection currents. This mechanism is thought to be at the origin of the magma erupted at midocean ridges.

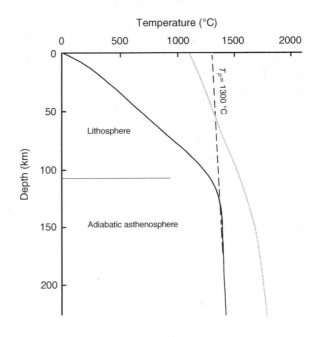

Fig. 4.1 Geotherm with a potential temperature $T_p = 1300\ °C$ and a kinematic viscosity of $4 \times 10^{15}\ m^2\ s^{-1}$, calculated for the oceanic domain of the Ionian microplate (from Pasquale et al. 2005). The grey curve is the garnet peridotite solidus by McKenzie and Bickle (1988)

There are various hypotheses concerning the cause of an increase in the mantle temperature. A first hypothesis involves that the occurrence of an instability in the outer core may form a protuberance, warmer than the surrounding mantle, which would then cause melting. This mechanism implies the ascent of magma in the mantle for about 2800 km and the formation of a single volcano or of a restricted volcanic area (hot spot) for many millions of years (see Sect. 1.4).

Another possibility is that the temperature rise may be caused by the heat released by the decay of uranium, thorium and potassium isotopes. When portions of the mantle melt, these elements concentrate in the magma. The mantle would therefore be strongly depleted, and thus it seems unlikely for the melting to occur due to radiogenic heating. However, the centers of the convective cells may be stagnant nuclei in which undepleted portions of the mantle remain undisturbed for a long time. This is yet another mechanism proposed to explain the origin of ascending magma columns in the mantle.

Another way to generate magma is to lower the mantle melting point by changing some physical–chemical properties. Such a mechanism may occur at the margins above a subducting plate (Fig. 4.2). The subducting plate carries with it a portion of crust (generally oceanic crust) and part of the overlying sediments, both rich in water. Due to subduction, these materials dehydrate and release water and other components in the mantle portion overlying the subducting plate. This causes a lowering of the melting point and the formation of magma with a particular chemical composition, typical of compressional zones.

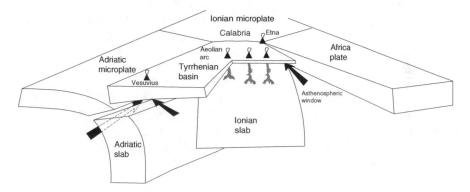

Fig. 4.2 Identification of the tectonic environments of formation of magma in the Central Mediterranean Sea subduction zone, where Ionian slab rollback opens the Tyrrhenian back-arc basin (Pasquale et al. 2005) (see Figs. 1.2 and 1.4). The isotopic ratios of the helium and carbon suggest that Mt Etna volcano is fed by the same type of mantle source as midocean ridge basalts, which in turn are results of the 'suction' of asthenospheric material from under the neighbouring Africa plate (Gvirtzman and Nur 1999). The existence of asthenospheric material under the Calabria crust explains not only the formation of Mt Etna, but also the recent uplift history (began about 700 kyr ago) of Calabria and of the Aeolian Island arc. Normal faulting and initiation of a new rift zone between Calabria and the Aeolian arc also took place during this time. The sinking Adriatic microplate creates also a local flow in the asthenosphere that produced the Quaternary magmatism at Mt Vesuvius and nearby

In metasomatic processes, a rock remains in the solid state and changes its chemical composition through the enrichment or depletion of chemical elements. As a final result, new minerals replace totally or in part the previous paragenesis. The mechanism is the same as metamorphism, but the fluid phase plays a more important role. The mantle is chemically heterogeneous even on a small scale. Depending on the content of some trace elements, such as rubidium, uranium, thorium, neodymium, and the isotopic ratios of strontium, neodymium and lead, a portion of the mantle is considered enriched or undepleted if its composition is different from the chondritic one. The transport of water and carbon dioxide, as well as alkali, radiogenic isotopes and heat, promotes the magma formation by increasing the temperature and lowering the mantle solidus. Until a few years ago, metasomatism and magma formation were considered to be very slow processes while today some researchers on the contrary believe that these phenomena may occur in a few tens or hundreds of years.

4.1.2 Melting by Rifting

In presence of a thermally anomalous asthenosphere, a pressure release in the mantle by adiabatical upwelling due to rifting processes can generate basaltic melts. Magma ascends to the Moho, where it cools, releases heat and solidifies forming new crust (underplating). This phenomenon could then lead to the formation of intrusions and volcanism. Magmatic underplating is an important feature of large igneous provinces. On the basis of basaltic compositions and mantle melts, Cox (1980, 1993) suggested that the ponding and fractionation of magmas at the base of the crust may be important processes in the generation of continental flood basalts; also magmatic underplating is a potentially large contribution to crustal accretion. This hypothesis has received supporting evidence in geophysical and geochemical observations for over 20 years (e.g. White and McKenzie 1989; Watts and ten Brink 1989; Caress et al. 1995; Kelemen and Holbrook 1995; Barton and White 1997; Farnetani et al. 1996; Bauer et al. 2000; Pasquale et al. 2003).

When underplating occurs in an extensional realm, it is possible to define a true stretching factor β given by

$$\beta = \frac{\beta_a h_c}{h_c - \beta_a h_b} \tag{4.1}$$

where β_a is the apparent stretching factor determined from the change in crustal thickness, which results from stretching when all the melt of thickness h_b is assumed to be added to the crust of initial thickness h_c. The volume of melt generated and the potential crustal addition will depend on the amount by which the mantle temperature exceeds its solidus and the melting behavior of the material involved. The mantle material can be assumed to be emplaced with a negligible heat loss and therefore an adiabatic rise is reasonable. This condition is verifiable if one considers the upwelling volume of an infinite cylinder of radius r_o embedded in an infinite medium in which the initial temperature is T_o for $0 \leq r \leq r_o$ and zero in the region $r > r_o$. The temperature as a function of the radius r and time t is

$$T(r,t) = \frac{T_o}{2\kappa t} \exp\left(-\frac{r^2}{4\kappa t}\right) \int_0^{r_o} \exp\left(\frac{r'^2}{4\kappa t}\right) I_o\left(\frac{r\,r'}{2\kappa t}\right) r'\,dr' \qquad (4.2)$$

where κ is the thermal diffusivity and I_o the modified Bessel function (Carslaw and Jaeger 1986; Furlong and Fountain 1986). This equation evaluated at the origin $(r = 0)$ becomes

$$T(0,t) = T_o\left[1 - \exp\left(-\frac{r_o^2}{4\kappa t}\right)\right] \qquad (4.3)$$

Figure 4.3 illustrates the minimum radius of the cylinder, calculated from (4.3), for which the material along the axis will cool by less than a specified fraction in a specified time. By considering that a 1 % cooling satisfies the assumption of an adiabatic rise, a minimum radius of about 40 km would be sufficient if the mantle material is emplaced at 3 cm yr^{-1}, which is a typical rate of motion for a plate, for an upwelling of 100 km.

A model of magma generation by extension must assume a mantle solidus and the function $M\,[T(z),\,z]$ describing the melting behavior of the material at different temperature T and depth z (see Pollack and Chapman 1977; Ahern and Turcotte 1979). The temperature and the melt fraction function for a set of assumed initial conditions is given by the equation on the energy content of the material undergoing partial melting

$$\rho\,c\,\theta(z) = \rho\,c\,T(z) + H_{fus}\,M[T(z),z] \qquad (4.4)$$

where $\theta(z)$ describes the potential temperature and $T(z)$ the actual temperature, ρ, c and H_{fus} are the density, the specific heat and the enthalpy of fusion, respectively. Equation (4.4) is nonlinear in the temperature and requires numerical calculations for its evaluation.

Fig. 4.3 Cylinder radius versus the rate of emplacement for different fractional cooling $(1-T/T_o)$. The emplacement velocity is given for an upwelling of 100 km

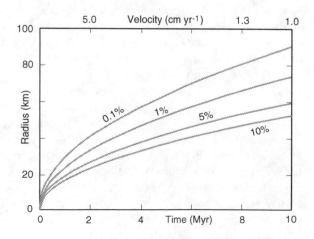

In the absence of melt, the geothermal flow q_o can be envisaged as a sum of a component q_r due to the radiogenic heat and a contribution q_t controlled by the rift time and extension amount, plus the heat flow q_a from the asthenosphere (see Sect. 3.2)

$$q_o = q_r + q_t + q_a \qquad (4.5)$$

The heat flow q_r can be estimated by means of (3.15) whereas the transient component q_t can be calculated from (3.16).

If melting occurs, the heat supplied by magma q_b contributes to the geothermal flow. By assuming that at distances greater than 1/4 of the thickness of the magmatic layer h_b the temperature may be modeled as a conductive cooling of the melt, the released heat flow is given by

$$
q_b = k \frac{T_b + \Delta T}{2\sqrt{\pi t \kappa}} \left[-2 \exp\left(-\frac{(h_1 + h_b)^2}{4\kappa t} \right) + 2 \exp\left(-\frac{h_1^2}{4\kappa t} \right) \right.
$$
$$
\left. + \exp\left(-\frac{(2h_2 + 2h_b)^2}{4\kappa t} \right) - \exp\left(-\frac{(2h_2 + h_1 + h_b)^2}{4\kappa t} \right) \right]
$$

$$(4.6)$$

where t is time and h_1 and h_2 are the thickness of the crust and of the lithospheric mantle above and below the underplated magmatic layer, respectively (Jaeger 1964). T_b is the temperature of the magma corrected by 1 mK m^{-1} for the liquid adiabatic upwelling in excess of what the temperature would have been in the region occupied by the magma layer if there were no melt, but accounting for the rise of the lithospheric isotherms caused by rifting. The temperature $\Delta T = 500$ °C is added to account for the fusion enthalpy of magma as it solidifies.

4.1.3 A Case Study of Magmatic Underplating

Pasquale et al. (2003) demonstrate that new crust can have been generated in the Tyrrhenian basin by continental extension above a moderately heated astheno-sphere. The Calabrian–Sicilian margin presents an average geothermal flow value of 125 mW m^{-2}, which is decidedly higher than those of the other margins of the western Mediterranean (80–100 mW m^{-2}) (cfr Figs. 1.4 and 3.5). This is partly a consequence of the younger age of the extensional events occurred in that area.

The heat flow budget of the margin was evaluated by means of (4.5). The contribution q_r is about 20 mW m^{-2} and the transient component q_t was modeled according to (3.16). For $\beta_a = 2.0$, as suggested by seismic data, and an age of the extensional event of 7 Myr the sum of the two contributions $q_r + q_t$ yields 55 mW m^{-2}. The third contribution q_a can be assumed to be 35 mW m^{-2}, a representative value for the heat flow coming from below the crust of the central-western Mediterranean (Pasquale et al. 1995). In summary, the different heat–flow components give $q_o = 90$ mW m^{-2}. Even if we incorporate in this budget the heat flow which derives from the radiogenic heat of the lithosphere mantle, there is a deficit in the calculated geothermal flow of about 35 mW m^{-2}.

A possible explanation could lie in the additional contribution due to the emplacement of mafic magma at the base of the continental crust. Partial melting generated by extension strongly depends on the potential temperature. Pasquale et al. (2003) solved (4.4) for a potential temperature of 1300 °C at the base of a 115 thick lithosphere, corresponding to 0.9 of the mantle melting temperature. Moreover, they assumed an initial geotherm typical of a Variscan crust with geothermal flow of 60 mW m^{-2}. By using a value of $\beta_a = 2.0$ no melting occurs. Two other models with moderately increased basal temperatures (by 50 and 100 °C, respectively) were then tested. For a potential temperature of 1350 and 1400 °C, a melt underplating of thickness of about 0.5 and 5.0 km can occur, respectively. These two values of thickness correspond to true stretching factors of 2.1 and 3.0. Figure 4.4 shows the calculated degree of partial melting as a function of depth under different hypothesis of stretching and thermal conditions at the lithosphere–asthenosphere transition. For a thermally normal asthenosphere, no melting occurs unless stretching exceeds values of 2.0. At slightly higher temperatures, partial melting can occur even for smaller β values. By using (4.6) with $k = 3.2$ W m^{-1} K^{-1} and $\kappa = 10^{-6}$ m^2 s^{-1} a 3.4 km thick magma layer underplating the crust, implying a true stretching factor of 2.5, can yield the 35 mW m^{-2} required to balance the geothermal flow observed on the continental margin (Fig. 4.5).

The contribution of heat flow from melt cooling lasts several years after the rift episode and even for larger periods than the characteristic thermal time or the solidification time of the melt layers. The explanation is that heat in the magma is stored in the surrounding medium and is therefore available also after that the magma has solidified.

Fig. 4.4 Partial melting as a function of depth at various amounts of extension β (1.5, 2.0, 2.5, 3.0, 4.0 and 5.0) and for three different values of asthenosphere potential temperature (1300, 1350 and 1400 °C)

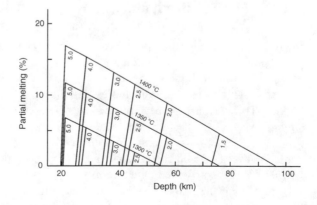

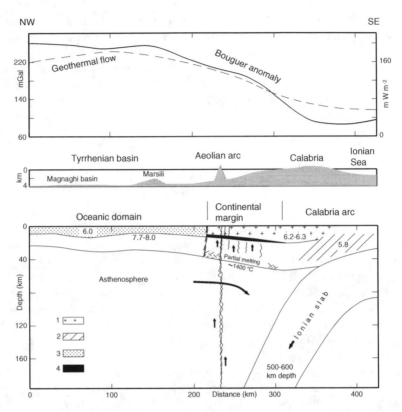

Fig. 4.5 Structural scheme of lithosphere–asthenosphere system of the Tyrrhenian basin (Fig. 1.4). Above, Bouguer gravity anomaly and geothermal flow pattern (after Pasquale et al. 1999). 1–continental crust, 2–low velocity layers, 3–oceanic crust, 4–crustal underplating. Values of velocity of compressional waves (in km s^{-1}) are indicated

4.2 Rheological Behavior

Magma is a heterogeneous physical–chemical system in which a silicate liquid phase prevails. Thus, conventionally, scientists refer to a homogeneous silicate magma. In silicate liquids, the anions O^{2-} are organized in tetrahedral fundamental structural units, coordinated essentially by Si^{4+}, which acts as a structure builder. The silicon may be replaced by other ions characterized by high field strength, such as B^{3+}, P^{5+} and Al^{3+}. SiO_4^{4+} tetrahedrons connect to each other and originate more or less complex polymers. The resulting structure, even in the simplest case of a liquid composition of SiO_2, is incomplete. The metal cations K^+, Na^+, Ca^{2+}, Mg^{2+} are called structure modifiers, as they tend to bond with oxygen by interrupting the -Si–O–Si–O chain, therefore generating weaker ionic bonds which result in a reduction of the point of polymerization of the system.

The complex structure of the silicate liquids is revealed by the physical properties that they assume under natural conditions. A peralkaline magma, with high concentrations of Na^+ and K^+ ions, will have low viscosity due to the effect of these ions in lowering the degree of melt polymerization. Basic melts, which have a higher ratio between oxygen atoms that do not form structural bridges and silicon, will have lower viscosity than acid magmas with same temperature and volatile content.

4.2.1 Temperature and Viscosity

Temperature, density and viscosity are the most important parameters which control the magma rheological behavior (McBirney and Murase 1984). Direct measurements of magma temperature are made by means of thermocouples in lava flows or through optical pyrometers in lava fountains. Most of the observations are related to basic magma of the Hawaiian Islands, that has temperatures in the range of 1050–1125 °C. Data for silica-rich lavas are scarce, because these kind of eruptions are rare; observations on Mount St. Helens lavas give an average value of about 850 °C. Temperatures of the main volcanic rocks observed during eruption are: 700–900 °C for rhyolite, 800–1100 °C for dacite, 950–1200 °C for andesite and 1000–1200 °C for basalt.

Experiments on magma were also carried out in the laboratory, by determining the temperature when melting starts (solidus) and ends (liquidus). For a material such as anhydrous garnet peridotite, i.e. a typical rock of the uppermost mantle, the following approximated relations are valid

$$T_S(z) = 3.0\, z + 1100 \quad \text{solidus} \tag{4.7}$$

$$T_L(z) = 1.5\, z + 1850 \quad \text{liquidus} \tag{4.8}$$

where z is in km and T in °C. By considering a linear relation between solidus and liquidus, the percentage of melt x_f at a given depth z and temperature $T_S(z) < T(z) < T_L(z)$ is given by

$$x_f = [T(z) - T_S(z)]/[(T_L(z) - T_S(z)] \tag{4.9}$$

This model is realistic in absence of a fluid phase (H_2O and CO_2). Figure 4.6 shows the dependence of the partial melt fraction of undepleted mantle (with 0.01 % H_2O) on temperature and depth and two typical geotherms for a continental and oceanic lithosphere. The solidus denotes a negative slope between 70 and 100 km, where the amphibole changes into pyroxene, olivine and garnet and releases water. The presence of CO_2 is less important, since its effect tends to compensate that of water.

The density of magma depends on its chemical composition, temperature and pressure. For natural melts, it varies from about 2200 kg m^{-3} for rhyolitic

Fig. 4.6 Dependence of the
partial melt fraction, in
percent, of undepleted mantle
(with 0.01 % water) on
temperature and depth
(dashed lines) (Ringwood
1975), and typical geotherms
with a potential temperature
T_p of 1300 °C (grey curves)

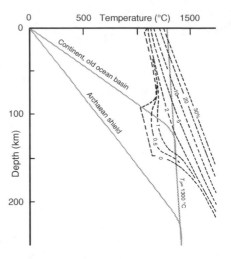

compositions, to 2800 kg m^{-3} for basaltic compositions. A method for estimating
density is based on the chemical composition

$$\rho = \sum x_i M_i / \sum x_i V_i \qquad (4.10)$$

where x_i, M_i, and V_i are the molar fraction, the molecular weight and the molar
volume of the various chemical species, respectively. The contrast in density
between the magma and the solid embedding is an essential factor of the rising
processes of natural melts. Differentiation processes related to solid–liquid sepa-
ration (fractional crystallization) will depend also on the difference in density.

A silicate liquid is rather viscous. Viscosity depends strongly on the chemical
composition, as well as temperature and pressure. In pure fluids, resistance to flow
is essentially caused by ionic or molecular cohesion; in magmas, it is complicated
by the presence of solids and gas bubbles. The Newton's relation between shear
stress σ and strain rate $d\varepsilon/dt$,

$$\sigma = \eta \frac{d\varepsilon}{dt} \qquad (4.11)$$

where η is the dynamic viscosity (a measure of the internal friction), is valid for
very fluid magmas at high temperature and with low concentration of crystals; it
can also be written in the form

$$\sigma = \sigma_0 + \eta \left(\frac{d\varepsilon}{dt}\right)^n \qquad (4.12)$$

in which σ_0 is the shear stress required for the flow onset. In Newtonian fluids,
$\sigma_0 = 0$ and $n = 1$. For a pseudoplastic magma, $\sigma_0 = 0$ and $n < 1$, while for a
plastic or Bingham fluid, σ_0 has a finite value and $n = 1$ (Fig. 4.7).

Fig. 4.7 Relationship
between shear stress and
strain rate

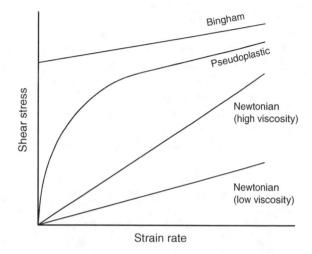

In absence of crystallized phases (liquidus), the dynamic viscosity variations
with temperature and pressure p can be expressed by

$$\eta = A_\eta \exp\left(\frac{E* + p\ V*}{RT}\right) \quad (4.13)$$

where $V*$ is the activation volume, R is gas constant and A_η is a constant. The
activation energy $E*$ is mainly dependent on the ratio $(Si + Al)/O$ of the melt. The
correlation is generally good, but in some cases it is negative, because, at high
pressures, it involves structural variations in silicate melts that alter the relationship
between builder elements and structure modifiers. In this regard, the role of Al^{3+},
whose octahedral coordination is strongly favored at high pressure, is significant.
This was found experimentally on giadeitic melts $(NaAl\ Si_2O_6)$ which undergo a
decrease in viscosity from about 3.5 kPa s at 0.5 GPa up to 0.5 kPa s at 2.5 GPa.

Fig. 4.8 Relationship
between dynamic viscosity,
η, temperature and
composition (a–pantellerite;
b–basic andesite; c–oceanic
island tholeiite; d–olivine
basalt; e–olivine
melanephelinite) at ambient
pressure

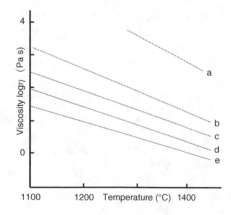

Figure 4.8 shows viscosity as a function of temperature, from laboratory results by Scarfe (1973) for some basic melts under ambient pressure.

Similarly to the role of the modifiers, the presence of volatiles dissolved in magma reduces viscosity, due to the depolymerizing effect of these substances. The magmatic gases have very complex and variable chemical compositions. The quantitatively dominant element is H_2O, along with a sometimes considerable amount of CO_2, while magmas are less rich in HCl, HF, SO_2, SO_3, S, N_2, rare gases, and NH_3. Volatiles are dissolved in the deep silicate melts (ipomagma), while in shallower conditions form a polyphase system (piromagma). The gas can leave the system forming a melt depleted of volatiles (epimagma).

As the effect of water on viscosity is more incisive in acid rather than in basic magmas (there are more Si-O bonds to break), a silica-rich magma, when erupted, will be more viscous than a basic one with an equivalent weight of water content at the same temperature. The other volatiles, such as HF, HCl and H_2S, act as structure modifiers following a pattern of hydrolysis similar to that of water. The behaviour of CO_2 is different, being its solubility at low pressures considerably lower. This is due to the low capacity of carbon dioxide to combine with oxygen to form $(CO_3)^{2-}$ in addition to the difficulty of being housed in molecular form in the melt structure. Its solubility increases in the presence of H_2O and in any case increases the polymerization (Burnham 1979).

Another factor that affects the magma rheology is the bubble content. The effect of bubbles on magma viscosity may be variable, depending on the degree of vesiculation, the size and distribution of bubbles, and the melt viscosity. In low viscosity melts, like basaltic melts, the volatile dissolution is of little influence on the overall viscosity, which is regulated by the temperature and composition. In contrast, a more acidic magma, at the outset, has high values of viscosity and can therefore be significantly influenced by volatile dissolution. A rhyolitic magma, for example, will have high viscosity, independently from the degree of vesiculation, unless it has an alkaline composition.

Also the presence of suspended crystals in magma increases viscosity; when, for example, the crystal content is about $60-65$ % of the total magma volume a sort of framework forms, due to the contact of granules, which considerably increases the internal friction, impeding the magma flow and ascent. The critical limit for the basaltic magma crystallization is estimated to be around 55 %, below which there is an effective potential of diffusion in the form of lava, and above which there is, instead, a potential to become intrusive bodies. In an acid magma, the critical limit is significantly lower. On the basis of experiments, the effective dynamic viscosity can be estimated from the relationship

$$\eta = \eta_L (1 - 1.67\, x_s)^{-2.5} \tag{4.14}$$

where x_s is the volume fraction of suspended solid and η_L is the viscosity of the liquid.

4.2.2 Lava Flows

Given the wide variety of physical states of the erupted material (which can range from completely liquid to completely solid), deformations can arise in several ways, from highly plastic (such as in lavas) to essentially rigid (e.g. in case of explosive fragmentation of a rhyolitic dome). As a consequence, volcanic materials have very variable mechanical properties. The rock strength decreases rapidly with the increase of the degree of partial melting; on the contrary, if magma crystallizes, its strength increases with the increase of the number of crystals. The greater strength of the material the greater is its tendency to oppose to explosive dismemberment. In this regard, fragmentation can be caused by tensile stresses or by shear stresses. The former may hold during the explosive bubble growth, when the gas pressure exceeds the tensile strength and the magma tension surface. The latter may occur, for example, when a vesicular magma is shattered by a shear stress, induced by a rather high sliding velocity, which exceeds the shear strength of the erupted mass.

The physical properties of magma and of the aggregates of pyroclastic and epiclastic particles have a large control over the nature of the resulting lava flows and clastic material. Viscosity, or the overall viscosity (for heterogeneous aggregates), is an important factor that can influence not only the mobility and the shape of the lava flows, but also the type of flow. In a laminar flow, the flow lines are practically parallel. In a turbulent flow, the flow lines are, instead, very irregular and dominated by vortices, resulting in a fluid mixing.

Experimentally, it is possible to determine a criterion to discern when the laminar flow of a Newtonian fluid is stable. For this purpose the Reynolds number R_e is used, which in terms of dynamic viscosity η is

$$R_e = \frac{\rho \, v \, D}{\eta} \tag{4.15}$$

where v is the mean velocity, ρ the density and D is the diameter of a circular conduit. For a channel with open flow, D is replaced with the hydraulic radius R (the ratio between the area of the fluid section and the length of the wet perimeter). According to (4.15), the transition between laminar and turbulent flow occurs at values of Re between 500 and 2000. For Bingham fluids, as lava and pyroclastic flows, the Reynolds criterion for turbulence is inadequate, because of the great strength and viscosity. For these fluids, the criterion for the turbulence is

$$R_e \geq 1000 \, B \tag{4.16}$$

where the number of Bingham B corresponds to

$$B = \frac{\tau_0 D}{\eta \, v} \tag{4.17}$$

with τ_o the mechanical strength of the material. By combining (4.15) and (4.17), the criterion of turbulence for Bingham fluid is given by

$$\frac{\rho \, v^2}{\tau_o} \geq 1000 \tag{4.18}$$

known as the Hampton number.

However, it is important to emphasize that the Reynolds number is inversely proportional to viscosity. For this reason, lava flows can take place in a laminar or turbulent way, but because of their high viscosity the laminar flow is more common. Only magma with low viscosity can have a turbulent flow, and this occurs only when the ground slope increases the already high velocity of the lava. Lava with high viscosity instead moves slowly, even on considerable slopes, due to its high threshold of internal strength. On the basis of experimental measurements on natural melts and artificial systems, empirical methods have been developed to calculate the viscosity from chemical composition and taking into account the temperature. Although these estimates are valid above the liquidus, it is still possible to have a quantitative indication of the viscosity contrasts between melts of different composition. An empirical relationship, relating viscosity to the ground inclination i, the thickness h and velocity v of the lava flow, is (Jeffreys 1925)

$$\eta = \frac{g \, \rho \, h^2 \operatorname{sen} i}{n \, v} \tag{4.19}$$

where g is the gravity acceleration, ρ is the density, and n varies from 3 for thick lava flows to 4 for thin lava flows.

4.3 Upwelling Mechanisms

It is not exactly known how much magma is produced at depth and how much is then erupted. Some rough estimates would indicate a ratio between 5 and 25. Among the mechanisms that may be responsible for the magma ascent, the most frequently invoked is buoyancy due to the lower magma density. The rise of magma in the crust can occur with a mechanism similar to the process of salt dome formation. Salt domes are formed by deposition of a low−density salt layer, originated by seawater evaporation. Then this layer is covered with sedimentary deposits with a high density. At shallow depth, the strength of the salt layer prevents the relative motions between the two masses of different density. As sedimentation proceeds, the salt layer moves at greater depths, and, consequently, at higher temperatures. The increase in temperature makes the salt layer plastic and facilitates the processes of deformation and upward flow of the lighter material. In volcanic processes, the low-density layer upon which magma formed permits the ascent of diapirs. If the ascent rate is high enough, the heat transferred by conduction to the embedded rocks is small and the magmatic body rises to the surface at a high temperature.

Another mechanism suggested to explain the rise of magma within the crust is the migration through fractures. Well−known examples are dykes, which can be observed at many eroded flanks of volcanoes. This mechanism assumes that magma pressure can generate fractures. Magma would then begin to fill the voids until the fracture reaches a characteristic length (critical length), which depends on the elastic properties of the medium, the fracture size and the density contrast between the fluid and the medium. Once the critical length, which may be several kilometer long, is attained the fracture begins to move upwards. While new cracks form in the direction of the surface, thus permitting the flow of magma, the pressure closes the lower gaps. The crack propagation can stop also at intermediate depths before reaching the surface.

Subsequently to this hypothesis, it was suggested that a fracture may propagate subcritically due to stress−corrosion, viscous damage and other mechanisms (Anderson and Grew 1977; Atkinson 1984; Rubin 1998). The opening of the fracture would result in a local decrease in pressure, which would then release through exhalation of the volatile content in magma. These gases would then act as an aggressive chemical agent in the upper part of the fracture reducing the energy required for fracturing the medium. The velocity of this mechanism is favored by the presence of water and by an increase in temperature, both of which reduce the stress needed to break the medium. An interesting result of the experimental studies on the mechanism of stress−corrosion is that the rock fracture, when it is done according to this mechanism, cannot generate noticeable seismic activity. This means that magma can rise to the surface without being accompanied by the typical earthquakes prior to its arrival. The temperature dependence of the viscoelastic properties of the host rock may play an important role in the energy balance during the slow, subcritical dyke propagation from a magma chamber. Recently, Chen and Jin (2011) showed that the viscous energy dissipation of the host rock could allow a short dyke slow grow (at a rate of 10^{-7}–10^{-5} m s^{-1}) under a modest over-pressure, and to accelerate when the intensity of the stress field close to the fracture toughness, followed by unstable dyke propagation.

The rate of propagation of dykes is controlled by the rate of the fracturing at the tip and by the flow rate of magma inside the dyke. When high energy is needed to fracture the host rock and magma viscosity is low, the rate of propagation is controlled by the rate of fracturing (fracture controlled regime). When the energy needed to fracture the host rock is low, propagation is controlled by the magma flow rate (magma−controlled regime). Maimon et al. (2012) studied the transition between these regimes in the case of a constant magma vesicularity and constant mass of gas in the cap. Pressure decrease during ascent leads to higher vesicularity and faster gas filtration through the magma and into a gas cap. A gradual increase of the mass of gas in the cap has an important role in accelerating the propagation rate of dykes.

When the magma moves to the surface, it yields a quantity of heat that gradually increases, because it encounters rocks with progressively lower temperatures. The amount of heat transferred depends on the difference in temperature between the magma and the surrounding rocks and on the mass and geometrical shape of the ascending magma. If magma reaches the surface, it means that it has retained enough heat to remain in a fluid state. If the ascent rate is too low, magma cools without reaching the surface. It has been calculated that the ascent rate of magma through fractures is several thousand times faster than that involving diapirism.

A practical method to estimate magma upwelling is based on the observation of variations in temperature, resistivity, density and magnetic properties of the uppermost crust, and the emissions and composition of gas. The increase of surface deformation, seismicity and tremors indicates magma motion through volcanic conduits. The occurrence of these physical and chemical anomalies forecast a recovery of the activity and the effects that an eruption may have on the environment. Analyzing and dating tephra and lava deposits allow to set out a volcano eruption pattern, with estimated cycles of intense activity and the size of eruptions. Satellites can measure the spread of an ash plume, as well as SO_2 emissions. Thermal imaging can monitor large, scarcely populated areas where it would be too expensive to maintain instruments on the ground.

4.4 Solidification and Cooling

4.4.1 Lava lake

Solidification processes involving phase changes are important in geology. The velocity at which the boundary between the solid and the liquid phase moves is the main point of interest, since the time necessary for a layer of a given thickness to solidify can be inferred from it. A general mathematical solution for this type of thermal conduction problem does not exist, but special explicit solutions. Here we use the Stefan problem to model the solidification of a lava lake, that is essentially the same of a freezing lake (see Baehr and Stephan 1998; Turcotte and Schubert 2002).

A lava lake is maintained at $z = 0$ at a constant temperature T_o which is lower than the solidification temperature T_m (Fig. 4.9). At the phase boundary $z = s$, which is moving downwards, the solidification of a layer of thickness ds releases enthalpy of fusion which flows towards the cooler upper part of the lake. The temperature $T = T(z, t)$ in the solidified lava satisfies the heat conduction equation

$$\frac{\partial T}{\partial t} = \kappa \frac{\partial^2 T}{\partial z^2} \tag{4.20}$$

with boundary and initial conditions

Fig. 4.9 Temperature profile
of the solidifying lava lake.
s is the distance between the
phase boundary and the
cooled surface at depth $z = 0$

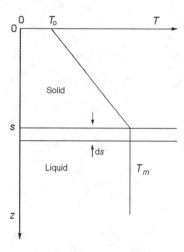

$$T = T_o \quad \text{for } z = 0 \text{ and } t > 0$$
$$T = T_m \quad \text{for } z = s \text{ and } t > 0$$
$$s = 0 \text{ for } t = 0$$

At the solidification boundary, the energy balance is

$$k \frac{\partial T}{\partial z} \, dt = H_{fus} \, \rho \, ds \tag{4.21}$$

where H_{fus} is the enthalpy of fusion, and the velocity of solidification is

$$\frac{ds}{dt} = \frac{k}{H_{fus} \, \rho} \left(\frac{\partial T}{\partial z} \right)_{z=s} \tag{4.22}$$

A solution of (4.20) is given by

$$T = T_o + C \, \text{erf}(\lambda) \tag{4.23}$$

where $\lambda = z(4\kappa t)^{-1/2}$. At the interface (4.23) becomes

$$T_m = T_o + C \, \text{erf}(\eta) \tag{4.24}$$

with the argument of the error function $\eta = s(4\kappa t)^{-1/2}$. Since η must be constant
and independent of t, and the increase in thickness of the solidified lava is proportional to $t^{1/2}$, we have

$$s = \eta \sqrt{4\kappa t} \tag{4.25}$$

By differentiating, we find the solidification velocity

$$\frac{ds}{dt} = \eta \sqrt{\frac{\kappa}{t}} \tag{4.26}$$

By combining (4.23) and (4.24) we obtain the dimensionless temperature ϑ, given by the ratio $(T-T_o)/(T_m-T_o)$ in the solidified lava $(0 \le z \le s)$ as

$$\vartheta = \frac{T - T_o}{T_m - T_o} = \frac{\mathrm{erf}(\lambda)}{\mathrm{erf}(\eta)} = \frac{\mathrm{erf}\left(\frac{\eta z}{s}\right)}{\mathrm{erf}(\eta)} \tag{4.27}$$

and the temperature gradient at $z = s$ as

$$\left(\frac{\partial T}{\partial z}\right)_{z=s} = \frac{(T_m - T_o)}{\sqrt{4\kappa t}\,\mathrm{erf}(\eta)}\frac{2}{\sqrt{\pi}}e^{-\eta^2} \tag{4.28}$$

By substituting (4.26) and (4.28) into (4.22), we get

$$\frac{H_{fus}\sqrt{\pi}}{c\,(T_m - T_o)} = \frac{e^{-\eta^2}}{\eta\,\mathrm{erf}(\eta)} = f(\eta) \tag{4.29}$$

a transcendental equation for determining η, where $c = k/(\kappa\rho)$ is the specific heat. Some values of $f(\eta)$ are given in Table 4.1. Alternatively, given a numerical value of the left side of (4.29), η can be found by iteratively calculating the right side of equation until a satisfying result is found.

As an example, we calculate the thickness of a solidified layer on a lava lake 60 days since its formation. The surface temperature is 20 °C and the lava melting temperature is 1100 °C, $H_{fus} = 400$ kJ kg^{-1}, $c = 1$ kJ kg^{-1} K^{-1}, $\kappa = 30$ m^2 yr^{-1}. From (4.29), we find $f(\eta) = 0.656$ and, from Table 4.1, $\eta = 0.884$. As $\eta = s(4\kappa t)^{-1/2}$ the thickness s is 3.9 m.

4.4.2 Intrusive Igneous Bodies

Magma which intrudes into the crust forms bodies of various shapes. Some are very massive (laccolith), others tabular (dyke, sill) and others more or less cylindrical (pipe) or spherical (magma chamber). Cooling of a magma body can be

Table 4.1 Values of the function $f(\eta)$

η	$f(\eta)$	η	$f(\eta)$	η	$f(\eta)$	η	$f(\eta)$
0.100	88.033	0.450	3.817	0.800	0.888	1.150	0.259
0.150	38.801	0.500	2.993	0.850	0.741	1.200	0.217
0.200	21.571	0.550	2.385	0.900	0.620	1.250	0.182
0.250	13.599	0.600	1.926	0.950	0.520	1.300	0.152
0.300	9.270	0.650	1.571	1.000	0.437	1.350	0.127
0.350	6.663	0.700	1.291	1.050	0.367	1.400	0.106
0.400	4.973	0.750	1.068	1.100	0.308	1.450	0.088

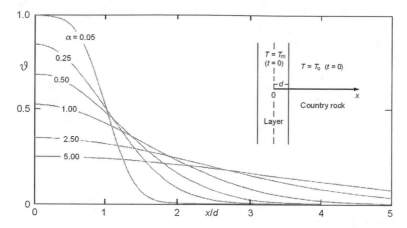

Fig. 4.10 $\vartheta = (T-T_o)/(T_m-T_o)$ as function x/d for an intrusive layer (dyke, sill) for different values of the Fourier number $\alpha = \kappa t/d^2$

approached by means of models with simple geometry (see e.g. Buntebarth 1984; Carslaw and Jaeger 1986), and also assuming that the thermal properties of the adjacent rock are the same of the intrusive body.

The cooling model of an intrusive layer of thickness $2d$ and the corresponding temperature field are illustrated in Fig. 4.10. As the thickness of the layer is smaller than its length, the temperature satisfies the one-dimensional, time-dependent heat conduction equation. By assuming that at $t = 0$ the temperature of the layer is T_m and of the rock wall is T_o, the temperature at a distance x from the center of the layer is

$$T(x,t) = T_o + \frac{T_m - T_o}{2}\left[\mathrm{erf}\left(\frac{x+d}{2\sqrt{\kappa t}}\right) - \mathrm{erf}\left(\frac{x-d}{2\sqrt{\kappa t}}\right)\right] \tag{4.30}$$

If the intrusion is of a spherical shape with radius R, the temperature at a distance r from the sphere center is described by

$$T(r,t) = T_o + \frac{T_m - T_o}{2}\left\{2\sqrt{\frac{t\kappa}{\pi r^2}}\left[\exp\left(-\frac{(r+R)^2}{4\kappa t}\right) - \exp\left(-\frac{(r-R)^2}{4\kappa t}\right)\right]\right.$$
$$\left. + \mathrm{erf}\left(\frac{r+R}{2\sqrt{\kappa t}}\right) - \mathrm{erf}\left(\frac{r-R}{2\sqrt{\kappa t}}\right)\right\}$$
$$\tag{4.31}$$

where T_o is the initial temperature of the adjacent rock. Figure 4.11 shows $\vartheta = (T-T_o)/(T_m-T_o)$ against r/R for different values of the Fourier number $\alpha = \kappa t/R^2$. Because of its finite dimension, a sphere cools more quickly than a layer.

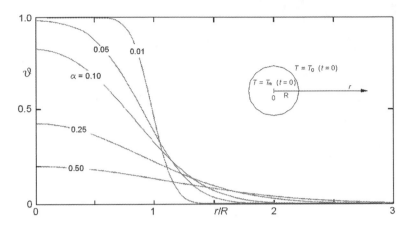

Fig. 4.11 $\vartheta = (T-T_o)/(T_m-T_o)$ as function r/R for a spherical intrusion (magma chamber) at different $\alpha = \kappa t/R^2$

4.4.3 Lava Covers

Lava covers, hydrothermal fluids and surface water migrating along a fault zone vary the temperature field in the nearby rocks. Such geological events can be modeled by the heating or cooling of a half space. If the half space, defined by $z > 0$, has at $t = 0$ a temperature T_1 and for $t > 0$ its boundary surface $z = 0$ is maintained at temperature T_{bs}, the temperature, as a function of position and time, is given by

$$T(z,t) = T_1 + (T_{bs} - T_1)\left[1 - \mathrm{erf}\left(\frac{z}{2\sqrt{\kappa t}}\right)\right] \tag{4.32}$$

Heat is transferred into the half space when $T_{bs} > T_1$. If $T_1 > T_{bs}$ the half space cools and the heat−flow density q at the surface $z = 0$ can be calculated by differentiating (4.32) according to Fourier's law and evaluating the result at $z = 0$

$$q = -k\left(\frac{\partial T}{\partial z}\right)_{z=0} = k\left(T_{bs} - T_1\right)\frac{\partial}{\partial z}\left(\mathrm{erf}\frac{z}{2\sqrt{\kappa t}}\right)_{z=0} = \frac{k\left(T_{bs} - T_1\right)}{\sqrt{\pi \kappa t}} \tag{4.33}$$

In the model of a cooling lava cover, it is necessary to consider at $t = 0$ half space at temperature T_1 overlaid with a lava flow of thickness h, having temperature T_2. The lava surface ($z = 0$) is at a constant temperature $T_o = 0$. The temperature field along the direction z perpendicular to the surface is described by

$$T(z,t) = T_1 + \frac{T_2 - T_1}{2}\left[2\,\mathrm{erf}\left(\frac{z}{2\sqrt{\kappa t}}\right) - \mathrm{erf}(\frac{z-h}{2\sqrt{\kappa t}}) - \mathrm{erf}\left(\frac{z+h}{2\sqrt{\kappa t}}\right)\right] \tag{4.34}$$

Figure 4.12 shows $\vartheta = (T-T_1)/(T_2-T_1)$ versus the ratio z/h for different $\alpha = \kappa t/h^2$.

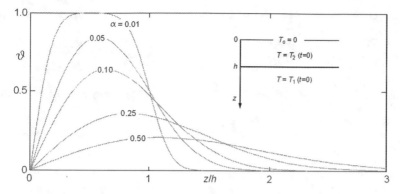

Fig. 4.12 $\vartheta = (T-T_1)/(T_2-T_1)$ as function z/h for a lava cover at different $\alpha = \kappa t/h^2$

All the temperature fields deduced from the foregoing analytical solutions are approximate, as thermal models do not consider some physical processes (magmatic pulses, magma collapse, hydrothermalism) and parameters that change in space and time (anisotropy, porosity, conductivity, specific heat, enthalpy of fusion). In order to overcome these limitations and to take into account complex geometries, numerical methods should be used (see Haenel et al. 1988; Baehr and Stephan 1998).

References

Ahern JL, Turcotte DL (1979) Magma migration beneath an ocean ridge. Earth Planet Sc Lett 45:115–122

Anderson OL, Grew PC (1977) Stress corrosion theory of crack propagation with applications to geophysics. Rev Geophys 15:77–103

Atkinson BK (1984) Subcritical crack growth in geological materials. J Geophys Res 89(B6):4077–4114

Baehr HD, Stephan K (1998) Heat and mass transfer. Springer, Berlin

Barton AJ, White RS (1997) Crustal structure of the Edoras Bank continental margin and mantle thermal anomalies beneath the North Atlantic. J Geophys Res 105:25829–25853

Bauer AJ, Neben S, Schreckenberger B, Emmermann R, Hinz K, Fechner N, Gohl K, Schulze A, Trumbull RB, Weber K (2000) Deep structure of the Namibia continental margin as derived from integrated geophysical studies: the MAMBA experiment. J Geophys Res 105:25829–25853

Buntebarth G (1984) Geothermics. Springer-Verlag, Berlin

Burnham CW (1979) Importance of volatile constituents. In: Yoder HD (ed) Evolution of igneous rocks. Princeton University Press, Princeton

Caress DV, McNutt MK, Detrick RS, Mutter JC (1995) Seismic imaging of hotspot-related crustal underplating beneath the Marquesas Islands. Nature 373:600–603

Carslaw HS, Jaeger JC (1986) Conduction of heat in solids, 2nd edn. Clarendon Press, Oxford

Chen Z, Jin ZH (2011) Subcritical dyke propagation in a host rock with temperature-dependent viscoelastic properties. Geophys J Int 186:1095–1103

Cox KG (1980) A model for flood basalt volcanism. J Petrol 21:629–650

Cox KG (1993) Continental magmatic underplating. R Soc London Philos Trans, Ser A 342:155–166

Farnetani CG, Richards MA, Ghiorso MS (1996) Petrological models of magma evolution and deep crustal structure beneath hotspots and flood basalts. Earth Planet Sci Lett 143:81–94

Furlong KP, Fountain DM (1986) Continental crustal underplating: thermal considerations and seismic-petrologic consequences. J Geophys Res 91:8285–8294

Gvirtzman Z, Nur A (1999) The formation of Mount Etna as the consequence of slab rollback. Nature 401:782–785

Haenel R, Rybach L, Stegena L (1988) Fundamentals of geothermics. In: Haenel R, Rybach L, Stegena L (eds) Handbook of terrestrial heat-flow density determination. Kluwer Academic Publishers, Dordrecht

Jaeger JC (1964) Thermal effects of intrusions. Rev Geophys 2:443–466

Jeffreys H (1925) The flow of water in an inclined channel of rectangular section. Phil Mag 49:793–807

Kelemen P, Holbrook S (1995) Origin of thick, high-velocity igneous crust along the U. S. East Coast Margin. J Geophys Res 100:10077–10094

Maimon O, Lyakhovsky V, Melnik O, Navon O (2012) The propagation of a dyke driven by gas-saturated magma. Geophys J Int 189:956–966

McBirney AR, Murase T (1984) Rheological properties of magmas. Annu Rev Earth Planet Sci 12:337–357

McKenzie D, Bickle MJ (1988) The volume and composition of melt generated by extension of the lithosphere. J Petrol 29:625–679

Pasquale V, Verdoya M, Chiozzi P (1995) On the heat flux related to stretching in the NW-Mediterranean continental margins. Studia Geoph Geod 39:389–404

Pasquale V, Verdoya M, Chiozzi P (1999) Thermal state and deep earthquakes in the Southern Tyrrhenian. Tectonophysics 306:435–448

Pasquale V, Verdoya M, Chiozzi P (2003) Heat-flux budget in the southeastern continental margin of the Tyrrhenian basin. Phys Chem Earth 28:407–420

Pasquale V, Verdoya M, Chiozzi P (2005) Thermal structure of the ionian slab. Pure Appl Geophys 162:967–986

Pollack HV, Chapman DS (1977) On the regional variation of heat flow, geotherms and lithospheric thickness. Tectonophysics 38:279–296

Ringwood AE (1975) Composition and petrology of the earth'mantle. McGraw-Hill, New York

Rubin AM (1998) Dike ascent in partially molten rock. J Geophys Res 103(B9):20901–20919

Scarfe CM (1973) Viscosity of basic magmas at varying pressure. Nature 241:101–102

Turcotte D, Schubert GL (2002) Geodynamics–application of continuum physics to geological problems, 2nd edn. Cambridge University Press, Cambridge

Watts AB, ten Brink US (1989) Crustal structure, flexure and subsidence history of the Hawaiian Islands. J Geophys Res 94:10473–10500

White RS, McKenzie D (1989) Magmatism at rift zones: the generation of volcanic continental margins and flood basalts. J Geophys Res 94:7685–7729

Zoback MD, Townend J (2001) Implications of hydrostatic pore pressure and high crustal strength for the deformation of intraplate lithosphere. Tectonophysics 336:19–30

Chapter 5
Heat in the Groundwater Flow

Abstract The presence of groundwater flow implies other heat transfer mechanisms rather than pure conduction. Several strategies have been developed to explore the heat transport associated with water flow. This chapter presents some analytical methods and shows how subsurface temperatures can provide a quantitative tool for inferring water flow in permeable layers and heat advection in hydrothermal systems. Thermal convection in deep aquifers and its potential are then analyzed by means of the dimensionless Rayleigh number.

Keywords Heat and water flow · Heat advection · Thermal convection · Borehole thermal logs · Hydrothermal systems · Deep carbonate reservoir

5.1 Background

Although heat conduction is dominating within the outer layers of the crust, water flow in the permeable formations implies heat to be transported also through mass motion. Such a mechanism is of strategic importance with regard to the availability and exploitation of geothermal energy. Complex heat and groundwater flow problems, for which non explicit solutions exist or can only be obtained with great effort, are preferentially solved by means of numerical methods (e.g. Haenel et al. 1988; Beck et al. 1989; Anderson and Woessner 1992; Swanson and Bahr 2004). However, analytical methods can be adequate to obtain information on coupled heat transport and single—phase water flow in saturated aquifers of mid—low permeability and simple geometry (see e.g. Anderson 2005; Pasquale et al. 2011a).

The basic theory of water displacement in the underground is similar to other diffusion theories. The driving force, or hydraulic potential, is given by pressure and the resistance is due to water viscosity and type of interstices through which the water flows. In many cases, groundwater flow can be described by Darcy's law, in which the velocity of water u is connected with its pressure in the form

V. Pasquale et al., *Geothermics*, SpringerBriefs in Earth Sciences,
DOI: 10.1007/978-3-319-02511-7_5, © The Author(s) 2014

$$u = -\frac{\kappa_p}{\eta_w} \nabla (p + \rho_w g z) = -\frac{\kappa_p \rho_w g}{\eta_w} \nabla H_h = -\lambda_c \nabla H_h \qquad (5.1)$$

where κ_p is permeability, η_w and ρ_w are the water dynamic viscosity and density, respectively, p is pressure, $H_h = z + p/(\rho_w g)$ is the hydraulic head, z is the elevation above a standard datum, g is the acceleration due to gravity, gH_h is hydraulic potential and $\lambda_c = \kappa_p \rho_w g/\eta_w$ is the hydraulic conductivity. The velocity u is the volumetric flow rate of water per unit area and it is referred to as the Darcy velocity. However, since the pores and cracks occupy only a small fraction of this area, u is not the actual water velocity, but the average velocity per unit area.

The energy equation governing convective heat transfer in permeable layers is discussed in several textbooks (e.g. Goguel 1976; Jessop 1990; Turcotte and Schubert 2002). Analytical solutions for aquifers were given by Stallman (1963 and 1965), Bredehoeft and Papadopulos (1965), Lubimova et al. (1965), Lachenbruch and Sass (1977) and Mansure and Reiter (1979). The heat transferred from rock to water depends on u, the volumetric heat capacity of water $\rho_w c_w$ and the thermal gradient ∇T, in the direction of u. The heat equation is thus expressed as

$$\rho c \frac{\partial T}{\partial t} = k \nabla^2 T - \rho_w c_w (\mathbf{u} \cdot \nabla T) \qquad (5.2)$$

where ρ_w and c_w are the density and specific heat of water, c, ρ and k are specific heat, density and thermal conductivity of the water–rock matrix, respectively, and t is time. In (5.2) thermal conductivity is assumed to be isotropic and temperature–independent, the thermal equilibrium to hold between the rock and water, and heat sources are neglected.

Flow in an aquifer may be caused by an externally applied pressure gradient. Under this condition, the heat transfer mechanism is known as advection. When the flow occurs due to buoyancy effect (i.e. a change in density caused by a thermal gradient), the mechanism is called thermal convection. If the horizontal thermal gradient is negligible and heat and water flows are steady, substitution of (5.1) into (5.2) gives

$$\frac{d^2 T}{dz^2} = \frac{\rho_w c_w \kappa_p}{k \eta_w} \left[\frac{d}{dz} (p + \rho_w g z) + g z \frac{d\rho_w}{dz} \right] \frac{dT}{dz} \qquad (5.3)$$

where ρ_w is a function of depth z. By introducing in this equation the hydraulic head and the expansion coefficient of water $\alpha_w = (d\rho_w/dT)/\rho_w$, it is obtained

$$\frac{d^2 T}{dz^2} = \frac{\rho_w^2 c_w g \kappa_p}{k \eta_w} \left(\frac{dH_h}{dz} + \alpha_w z \frac{dT}{dz} \right) \frac{dT}{dz} \qquad (5.4)$$

This equation contains terms expressing the conductive heat transfer, heat advection and thermal convection in a homogenous permeable layer. In case of heat advection, the buoyancy forces are assumed to be negligible, whereas in a thermal convection system water motion is entirely due to buoyancy.

5.2 Heat Advection

5.2.1 Vertical Flow

The water flows from recharge areas, where precipitation seeps downwards from the ground surface and reaches the saturated zone, to discharge areas, where it is discharged to streams, lakes, ponds or swamps, is an additional mechanism of heat transfer which sums to pure conduction. For a combined conductive and advective heat transfer, (5.4) becomes

$$\frac{d^2 T}{dz^2} - \frac{c_w \rho_w u_z}{k} \frac{dT}{dz} = 0 \tag{5.5}$$

where u_z is the Darcy velocity in the vertical direction (positive downwards). The solution of (5.5) is

$$T = T_1 + (T_2 - T_1) \frac{\exp\left(\beta_z z/h\right) - 1}{\exp\left(\beta_z\right) - 1} \tag{5.6}$$

where $\beta_z = c_w \rho_w u_z h/k$ is a dimensionless parameter, which may be positive or negative depending on whether u_z is downward or upward, and T_1 and T_2 are the temperatures at the top and at the bottom of the vertical distance h within the aquifer (Fig. 5.1a). Equation (5.6) is valid for steady-state thermal conditions and for a uniform, isotropic, homogeneous and saturated aquifer. The flow rate is assumed to be constant and small enough so that thermal equilibrium is maintained between the water and the rock matrix.

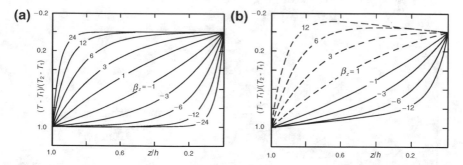

Fig. 5.1 Ratio $(T-T_1)/(T_2-T_1)$ versus z/h. **a** for different vertical water flows in recharge (β_z) or discharge ($-\beta_z$) areas; **b** for $(\Gamma_o\beta_o)/(\Gamma_z\beta_z) = 0.2$ (dotted curve) and $(\Gamma_o\beta_o)/(\Gamma_z\beta_z) = -0.2$ (solid curve) for different β_z

5.2.2 Two-Dimensional Flow

Because most layers are sloping and because surface topographic relief usually exists across an aquifer, heat and water flow, particularly in semi-confining layers, is neither purely horizontal nor purely vertical (Fig. 5.2). In this case, Lu and Ge (1996) demonstrated that the solution is an extension of (5.5). Assuming a linear variation of temperature along the horizontal direction, in the left–hand member of the equation we must add the term $-\Gamma_o(\beta_o/h)$ that accounts for the constant horizontal flow of heat and water. Therefore we have

$$\frac{\mathrm{d}^2T}{\mathrm{d}z^2} - \frac{\beta_z}{h}\frac{\mathrm{d}T}{\mathrm{d}z} - \frac{\beta_o}{h}\Gamma_o = 0 \tag{5.7}$$

where $\beta_o = c_w\rho_w u_o h/k$, Γ_o and u_o are the horizontal components of the thermal gradient and the Darcy velocity, respectively. The gradient Γ_o is taken as positive (or negative) when the heat flow is leftward (or rightward). The solution of (5.7) is

$$T = T_1 + (T_2 - T_1)\left[\frac{\exp(\beta_z z/h) - 1}{\exp(\beta_z) - 1} + \frac{\Gamma_o\,\beta_o}{\Gamma_z\,\beta_z}\left(\frac{\exp(\beta_z z/h) - 1}{\exp(\beta_z) - 1} - \frac{z}{h}\right)\right] \tag{5.8}$$

where Γ_z is the vertical thermal gradient. In the absence of horizontal heat or water flow ($\Gamma_o = 0$ or $u_o = 0$), (5.8) reduces to (5.6). The sensitivity of $(T-T_1)/(T_2-T_1)$ to the ratio $(\Gamma_o\,\beta_o)/(\Gamma_z\,\beta_z)$ is shown in Fig. 5.1b.

Reiter (2001) suggested that the analysis of the thermal effect of groundwater can be practicable by comparing the thermal gradient Γ_z with temperature and depth. Integrating (5.7) once we obtain

$$\Gamma_z = \frac{\beta_z T}{h} + \frac{\beta_o\,\Gamma_o\,z}{h} + c \tag{5.9}$$

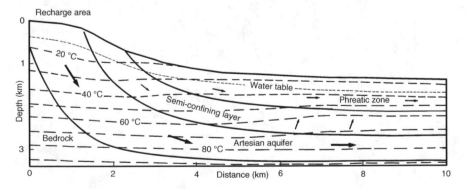

Fig. 5.2 Outline of an advectively disturbed thermal regime in a typical aquifer system. Notice that isotherms are not always horizontal. Equipotential lines are perpendicular to water velocity which is indicated by arrows

where c is the integration constant. Equation (5.9) can be seen as a plane whose slopes in relation with the axes of temperature and depth contain information on the vertical and horizontal components of the Darcy velocity, respectively.

5.2.3 Péclet Number

The quantities β_z and β_o measure the relative efficiency of a permeable layer for the simultaneous heat transport by both water flow and pure conduction. They appear to be analogues of the thermal Péclet number, which quantifies the potential for advection to perturb the temperature−depth distribution. The vertical Péclet number Pe_z can be expressed as the ratio of the advected heat flow q_{ad} and the conducted heat flow q_c over a characteristic length L

$$Pe_z = \frac{q_{ad}}{q_c} = \frac{\rho_w c_w |u_z| (T_2 - T_1)}{k (T_2 - T_1)/L} = \frac{\rho_w c_w |u_z| L}{k} \tag{5.10}$$

When $Pe_z \gg 1$ vertical advection dominates, while the conductive component prevails for $Pe_z \ll 1$. The length L can be chosen in many ways and is often selected on the basis of the scale of the flow system (see e.g. Domenico and Palciauskas 1973; Clauser and Villinger 1990). Normally, it is assumed to correspond to the thickness of the aquifer. By substituting u_o for u_z into (5.10), it is possible to derive also the horizontal Péclet number.

5.2.4 Application to Borehole Thermal Logs

The foregoing analytical solutions for heat and water flow can be applied to borehole thermal logs, carried out by means of high precision equipment (e.g. uncertainty of the order of 0.01 °C). Figure 5.3 shows an example of a temperature profile recorded in a 100 m deep borehole. The thermal log is characterized by an evident upward concave profile, which may denote the flow of cold water. The aquifer bottom is at a depth of 70 m. Thermal conductivity measured on core specimens (arenaceous marl) is on the average 2.0 W m^{-1} K^{-1}.

The analysis procedure consists in matching temperature and thermal gradient data with theoretical curves obtained from (5.6), (5.8) and (5.9) (curves A, B and C, respectively, of Fig. 5.4). By re-arranging (5.6) into a simplified form, the curve A is obtained

$$T = a_1 + b_1 e^{c_1 z} \tag{5.11}$$

where $a_1 = T_1 - (T_2 - T_1)/ \left(e^{\beta_z} - 1 \right)$, $b_1 = (T_2 - T_1)/ \left(e^{\beta_z} - 1 \right)$ and $c_1 = \rho_w c_w u_z/k$. Curve B is obtained by rewriting (5.8) as

Fig. 5.3 Temperature T, thermal gradient Γ_z and stratigraphy in a hole at Acqui Terme, NW Italy (data in Pasquale et al. 2010)

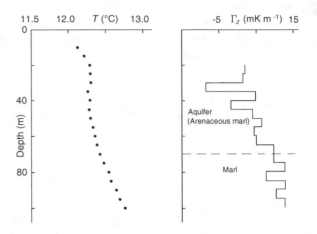

$$T = a_2 + b_2\, e^{c_2 z} - d_2\, z \tag{5.12}$$

where $a_2 = T_1 - (T_2 - T_1)(1 + \delta)/(e^{\beta_z} - 1)$, $b_2 = (T_2 - T_1)(1 + \delta)/(e^{\beta_z} - 1)$, $\delta = (\Gamma_o \beta_o)/(\Gamma_z \beta_z)$, $\delta = (\Gamma_o \beta_o)/(\Gamma_z \beta_z)$, $c_2 = c_w \rho_w u_k/k$ $d_2 = u_o \Gamma_o x/u_z$ and from (5.9) the curve C is

$$\Gamma_z = a_3 + b_3 z + c_3 T \tag{5.13}$$

with $a_3 = c$, $b_3 = c_w \rho_w u_o \Gamma_o/k$ and $c_3 = c_w \rho_w u_z/k$.

    Provided that the volumetric heat capacity of water, the average bulk thermal conductivity and the horizontal thermal gradient are known, the curve coefficients c_1, c_2 and c_3 thus contain information about u_z, whereas b_3 and d_2 give u_o. A least–square fitting procedure can be used to determine curve coefficients (see Reiter 2001; Verdoya et al. 2008; Pasquale et al. 2010, for details). After velocity is

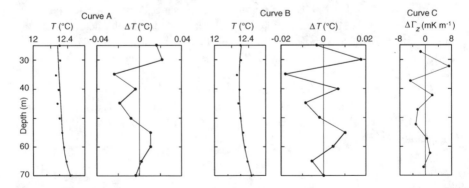

Fig. 5.4 Curve fit to thermal data of the borehole of Fig. 5.3. ΔT is the difference between observed and calculated temperature (curves A and B), and $\Delta\Gamma_z$ is the difference between observed and calculated thermal gradient (curve C) (see also Table 5.1). Calculations were carried out only for the section of the hole affected by water circulation (25–70 m)

Table 5.1 Darcy velocity, Péclet number and statistics of the curves used for fitting thermal data of Fig. 5.3

Depth range (m)	Curve	Darcy velocity (m s^{-1})		Péclet number		SSE*	RMSE**
		vert.	horiz.	vert.	horiz.		
25–70	A	4.12×10^{-8}	–	3.9	–	0.0048	0.0262
	B	0.21×10^{-8}	1.89×10^{-8}	0.2	1.8	0.0011	0.0125
	C	1.98×10^{-8}	1.47×10^{-8}	1.9	1.4	0.0001	0.0041

*SSE summed square of residual, ** RMSE root mean square error

determined and the characteristic length L is fixed, the Péclet number can be estimated.

Table 5.1 shows the values of the Darcy velocity vertical and horizontal components, and the Péclet number, as inferred from the curve coefficients. The water density and specific heat are assumed to be 1000 kg m^{-3} and 4186 J kg^{-1} K^{-1}. A horizontal thermal gradient value of 3.3 mK m^{-1} is assumed for calculating the horizontal component of the Darcy velocity. The goodness of fit of the three theoretical curves can be evaluated from the summed square of the residual and the root mean square error. The latter is a measure of the variation of the observed values around the calculated values.

The results show that the thermal profile is consistent with a slow downward flow of water with a vertical Darcy velocity varying from 0.2×10^{-8} to 4.1×10^{-8} m s^{-1}, i.e. from about only 0.06 to 1.3 m per year. Less variability is obtained for the horizontal component, which results in the range of 0.5–0.6 m per year. Of course, curve A cannot reveal any horizontal motion of water, as it assumes that horizontal thermal gradient is zero. The data analysis in terms of vertical thermal gradient indicates that the horizontal and vertical components of the Darcy velocity are comparable

It should be stressed that the underground thermal regime may contain a discernible climatic signal, as a consequence of the changes of ground surface temperature over the past few decades. This may cause a shift in the temperature–depth data that masks the effect of water. In this case, temperature data should be preliminarily treated for such a climatic noise (see e.g. Pasquale et al. 2000; Verdoya et al. 2007).

5.2.5 Hydrothermal Systems

Analytical solutions can also provide a quantitative tool for analyzing the water temperature along a deep flow path (Schoeller 1962; Pasquale et al. 2011a). Figure 5.5 shows a sketch of the horizontal branch of an aquifer, at depth H below the surface and of thickness h_o, in which water moves at a Darcy velocity u. The water temperature T varies only along the horizontal x axis. In a section dx of the aquifer, during time dt = dx/u, the heat per unit width that enters is $Q_a = q\mathrm{d}x^2/u$

whereas the heat flowing out is $Q_b = k(T-T_o)\mathrm{d}x^2/(H\,u)$, where q is heat–flow density beneath the aquifer, T_o the temperature at the surface and k the rock thermal conductivity, assumed to be uniform with depth. The gain of heat $Q = (Q_a-Q_b)$ is

$$Q = \left(\frac{q}{u} - k\frac{T-T_o}{H\,u}\right)\mathrm{d}x^2 \tag{5.14}$$

and the corresponding increase of water temperature is $\mathrm{d}T = Q/(\rho_w c_w h_o \sigma \mathrm{d}x)$, where σ is the effective porosity. Thus, along the direction x, we obtain

$$\frac{\mathrm{d}T}{\mathrm{d}x} = \left(q - k\,\frac{T-T_o}{H}\right)\frac{1}{\rho_w c_w \sigma h_o\, u} \tag{5.15}$$

If at $x = 0$ the temperature $T = T_i$, the normalized temperature ψ as a function of X (Fig. 5.5) is given by

$$\psi(X) = \exp\left(-\frac{X}{\alpha}\right) \tag{5.16}$$

where $\psi(X) = (T(x)-T_H)/(T_i-T_H)$, $X = x/H$ and T_H is the background temperature at depth H given by $T_o + qH/k$. The parameter $\alpha = c_w\rho_w u h_o\sigma/k$ is dimensionless. This approach assumes that water and the aquifer inner boundaries have the same temperature because of the large thermal resistance of the rocks, the slow water flow and the large contact surface.

For a descending aquifer, (5.16) becomes

$$\varphi(x) = \frac{\Gamma_z\, x}{\cot i + \alpha} \tag{5.17}$$

where $\varphi(x) = T(x)-T_i$ is the temperature of water along the horizontal x-axis, T_i is the temperature of water at the beginning of the flow path, Γ_z is the background

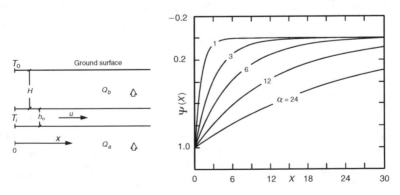

Fig. 5.5 Temperature $\psi(X)$ for different values of α along the horizontal branch of an aquifer (see text)

Fig. 5.6 Scheme of an
ascending branch of water
circuit (see text)

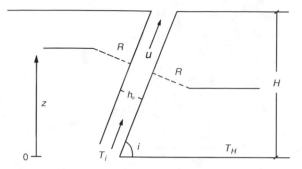

vertical thermal gradient and i is the slope of the aquifer. If a branch of a
hydrothermal circuit ascends (Fig. 5.6), the water temperature as a function of z is
given by

$$\vartheta\left(z\right) = -\Gamma_z z + (T_i - T_H)\exp\left(-\frac{2z}{R\alpha\sin i}\right)$$
$$+ \frac{\Gamma_z R\alpha\sin i}{2}\left[1 - \exp\left(-\frac{2z}{R\alpha\sin i}\right)\right] \tag{5.18}$$

where $\vartheta(z) = T(z) - T_H$ and T_H is the unperturbed temperature at depth H ($z = 0$).
The term $R = h_o e^5$ represents the thickness of the medium, surrounding the
aquifer and perturbed by the heat exchange.

The ascending branch of a hydrothermal circuit, from the reservoir to a spring,
can also be formed by a narrow, fracture zone (sub-vertical fault). If the flow rate
of the spring is high, the water ascent involves a negligible heat loss. An ascending
branch may be more or less irregular in shape and dimensions. For the sake of
simplicity, if we model the temperature decrease through a single sub-vertical
cylindrical conduit, by replacing h_o with the radius r of the conduit, (5.18)
becomes

$$\phi(z) = -\Gamma_z z + \frac{\Gamma_z \sin i}{A}\left[1 - \exp\left(-\frac{A z}{\sin i}\right)\right] \tag{5.19}$$

where $\phi(z) = T(z) - T_H$ is the water temperature as a function of depth and
$A = 2k/c_w\rho_w u\, r^2\ln(R/r)$.

By assuming that water heats as described in (5.16–5.19), the temperature in a
hydrothermal circuit can be modeled. Figure 5.7 and Table 5.2 summarize the
estimated temperature, together with the geometrical characteristics of the flow
path and the water travel time, in a hydrothermal system of NW Italy (Pasquale
et al. 2011a). The circuit recharge zone is located 14 km from the hot spring,
where the basement crops out. Meteoric water rapidly leaks down to a maximum
depth of 2800 m. For a Darcy velocity of 0.2 m day^{-1} and an aquifer of effective
thickness (the product of the saturated layer thickness h_o by effective porosity σ) of
1.0 m, the application of (5.17) gives a water temperature of 44.2 °C at this depth.

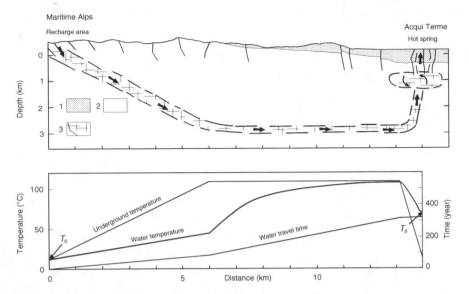

Fig. 7 Above: pattern of the Acqui Terme (NW Italy) hydrothermal system and main geological structures along water flow path. 1: Oligo–Miocene sedimentary sequences; 2: pre-Oligocenic basement; and 3: fractured zone and faults. Below: underground temperature, water temperature and water travel time. $T_o = 12.5$ and $T_s = 70\ °C$ are the water temperatures in the recharge and discharge zones, respectively (modified after Pasquale et al. 2011a)

Table 5.2 Hydraulic parameters, temperatures and water travel time in the four branches of the Acqui Terme hydrothermal system (NW Italy)

Aquifer branch	Horizontal extent (m)	Slope value	Effective thickness* (m)	Darcy velocity (m day^{-1})	Parameter α	Water temperature (°C)		Water time (years)
						in	out	
I–descending	6000	25°	1.0	0.2	4.4	12.5	44.2	90.6
II–horizontal	7150	0°	0.3	0.1	0.7	44.2	107.9	223.1
III–ascending	700	70°	0.3	7.1	46.9	107.9	70.9	0.8
VI–conduit of spring	150	80°	0.17**	0.1 (m s^{-1})	–	70.9	70.0	2.4 (h)

*Product between thickness and interconnected porosity of the aquifer, ** Cross–sectional radius

   Most of the heat is absorbed along the deepest, horizontal branch of the aquifer. Since a large fraction of the system recharge might be widely dispersed and porosity should decrease with depth, a smaller effective thickness of aquifer and a lower water velocity (about 0.3 and 0.1 m day^{-1}, respectively) is hypothesized. At the end of this branch, over a distance of about 7 km, according to (5.16), the water temperature increases to about 108 °C, i.e. the maximum temperature of groundwater estimated from geochemical data.

The ascending section includes a deeper branch: a first part, rising at an angle of 70°, from 2800 m to a reservoir located at intermediate depth. The upward water flow might be fast and occur through fractures with an increased porosity. The reservoir depth and temperature are inferred by extrapolating the thermal gradient observed in holes drilled in the metamorphic basement. For a thermal gradient of 70 mK m^{-1} the depth of the reservoir appears to be of about 850 m. Assuming an effective thickness of 0.3 m and a water velocity of 7.1 m day^{-1}, the application of (5.18) yields a decrease of 37 °C when water enters the reservoir, i.e. to 71 °C. The main spring discharges water at 70 °C and at a constant rate of 9 dm^3 s^{-1}. This implies that the water final ascent occurs at a velocity of 0.1 m s^{-1} through a conduit of a radius of only 0.17 m. The water residence time within the hydro-thermal circuit is of the order of 315 years.

5.3 Thermal Convection

Thermal convection is considered to be a major mechanism for mass transfer in many geological environments, spanning from sedimentary basins (see e.g. Aziz et al. 1973; Wood and Hewett 1982; Haenel et al. 1988; Domenico and Schwartz 1998; Phillips 1991; Raffensperger and Vlassopoulos 1999; Pestov 2000; Anderson 2005; and references therein), to geothermal fields (see Hanano 1998) and fractured rocks (Murphy 1979; Zhao et al. 2003; Kühn et al. 2006; Pasquale et al. 2013). The potential of this mechanism can be evaluated by means of the Rayleigh number analysis.

5.3.1 Rayleigh Number

Thermal convection may occur under appropriate heat sources and permeability. Convection in an aquifer heated from below can be analyzed with the Rayleigh number, Ra, defined as (Horton and Rogers 1945; Lapwood 1948)

$$Ra = \frac{g\,\alpha_w \rho_w c_{pw}\, \kappa_p\, L\, \Delta T}{v_w\, k} \tag{5.20}$$

where g is acceleration due to gravity, α_w, v_w and $\rho_w c_{pw}$ are the expansion coefficient, kinematic viscosity and volumetric heat capacity, respectively, of the saturating water, L is the length over which convection occurs, ΔT is the temperature difference between the top and bottom of this length interval, κ_p and k are the permeability and the thermal conductivity of the aquifer.

In the temperature range expected within an aquifer, v_w can be evaluated by (Holzbecher 1998)

$$v_w = 10^{-3}[1 + 0.015512(T - 20)]^{-1.572} \quad 40\,°C < T < 100\,°C \qquad (5.21)$$

$$v_w = \rho_w 0.2414 \ 10^{\frac{247.8}{T+133.15}} \ 10^{-4} \quad 100\,°C < T < 300\,°C \qquad (5.22)$$

where water density ρ_w varies with temperature T as

$$\rho_w = 1000 \left[1 - \frac{(T - 3.9863^2) \ (T + 288.9414)}{508929.2 \ (T + 68.12963)} \right] \qquad (5.23)$$

Thermal conductivity k can be calculated with (2.49), known the porosity, the thermal conductivity of the water and the solid matrix; it must be reduced to in situ conditions by taking into account the change in porosity with depth due to compaction and the temperature effect (see Sect. 2.4.2).

In practical terms, the Rayleigh number helps to differentiate between purely conductive systems, where isothermal constant density models are applicable, and low−grade hydrothermal systems, where non−isothermal models must be used. When the Rayleigh number exceeds the critical value of $4\pi^2$, convection starts and water flow forms convective cells. Such a flow is often termed as density−driven convection to stress the role of density variations in driving the motion (e.g. Holzbecher 1998; Turcotte and Schubert 2002).

5.3.2 Deep Carbonate Reservoirs

Deep carbonate aquifers are probably the most important thermal water reservoirs outside of volcanic areas. A review of the recent knowledge on thermal water resources in deep carbonate units can be found in the paper by Goldscheider et al. (2010). In these reservoirs, water is connected to regional flow systems, characterized by cross-formational hydraulic continuity (Tóth 1995; Frumkin and Gvirtzman 2006). Groundwater flow is primarily gravity−driven and is caused by the topographic gradient, but other phenomena, such as sediment compaction, tectonic compression, density differences in the water permeating rocks and thermal convection may act as additional driving forces.

An example for thermal convection in a deep carbonate aquifer as inferred from geothermal data is given by Pasquale et al. (2013). They analysed, through an inversion technique, temperature data from hydrocarbon exploration wells available in the east Po Plain (Italy). The obtained thermal gradient is quite low within the deep carbonate layer (14 mK m^{-1}), while it is larger (53 Mk m^{-1}) in the overlying low−permeability formations. In the shallower formations, the thermal gradient is close to the regional average (21 mK m^{-1}) (Fig. 5.8). As the thermal conductivity variation is small within the sedimentary sequence, this vertical change argues for convective processes occurring in the deep carbonate layer. Since the hydrogeological characteristics (including litho−stratigraphic sequences and structural setting) hardly permit advection, thermal convection could be the

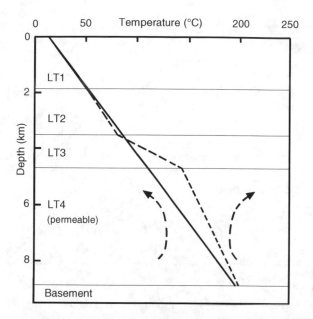

Fig. 5.8 Temperature versus depth as inferred from temperatures recorded in petroleum wells (broken line) and by extrapolating the regional thermal gradient (continuous line) in the east of the Po Plain (Italy). Arrows indicate likely thermal convection in the permeable carbonate reservoir. Lithologic units: LT1—Quaternary sands and alluvium and Pliocene marine clays and sandy clays, LT2—marls and arenaceous marl of Miocene age, LT3—manly formed by calcareous and argillaceous marls of Palaeogene age and LT4—Mesozoic carbonates

driving mechanism of circulation. Convection in the deep carbonate reservoir causes a decrease of thermal gradient and a significant increase in the overlying low−permeability unit.

The potential for thermal convection to take place within the carbonate reservoir was quantified with (5.20), by assuming that the parameters are independent from pressure and regarding the thickness of the reservoir as the length L of convection. The porosity change with depth was evaluated with an empirical relation (Pasquale et al. 2011b)

$$\phi = 0.180 \, e^{-0.396z} \tag{5.24}$$

where the depth z is expressed in km. The thermal conductivity of the carbonate rock matrix was assumed to be 4.0 W m^{-1}K^{-1}. By taking $Ra = 4\pi^2$, $g = 9.8$ m s^2 and values of $\rho_w c_{pw} = 4032$ kJ m^{-3} K^{-1}, $\alpha_w = 0.82 \times 10^{-3}$ K^{-1} and $v_w = 2.60 \times 10^{-7}$ m^2 s^{-1}, (5.20) becomes

$$\left(\frac{dT}{dz}\right)_{min} = \frac{10.0 \times 10^{-10}}{\kappa_p L^2} \tag{5.25}$$

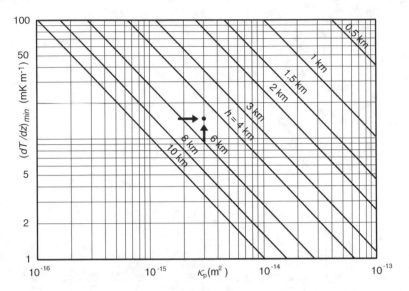

Fig. 5.9 Thermal gradient $(dT/dz)_{min}$ as a function of the permeability κ_p and thickness L of the carbonate reservoir in the east of the Po Plain (Italy). The full circle indicates the average regional value

which provides the minimum thermal gradient required for convection to occur in the reservoir. Figure 5.9 shows $(dT/dz)_{min}$ as a function of permeability k and thickness L. By considering the thermal gradient (14 mK m^{-1}) and the thickness (about 5 km) of the Po Plain carbonate aquifer, a permeability larger than 3.0×10^{-15} m^2 is required for thermal convection.

This value of permeability is consistent with the minimum permeability range of 5×10^{-17}–10^{-15} m^2 required for thermal convection to occur (Forster and Smith 1989; Manning and Ingebritsen 1999). Permeability depends on several parameters, such as the length to width ratio (or aspect ratio) of the flow domain and the presence of faults or other discontinuities (Smith and Chapman 1983; Clauser and Villinger 1990). Compared with the other sedimentary rocks, carbonates exhibit a wide variety of vertical and horizontal heterogeneities. In general, during post-depositional processes, the pores of the carbonates can be affected by mineral dissolution, re-crystallization or replacement by other minerals. Primary porosity is often poor and it is the secondary porosity (e.g. by fissure/ fracture) of tectonic origin or related to unloading processes which is more important. In carbonate reservoirs, water flow might occur through micro—and macro–fractures, which might have been broadened by karst phenomena. Finally, it must be kept in mind that a permeable layer may start convecting even at a Rayleigh number lower than the critical one, i.e. when the bottom boundary temperature of the layer is much higher than that of the top boundary (Garg and Kassoy 1981). This condition can be described by the over-heat ratio, i.e. a ratio of the difference in temperature between the top and bottom boundaries and temperature of the bottom boundary (see Hanano 1998). For the Po Plain carbonate reservoir, the over-heat ratio is on average of 0.45 and is therefore compatible with thermal convection.

References

Anderson MP (2005) Heat as a ground water tracer. Ground Water 43:951–968
Anderson MP, Woessner WW (1992) The role of the postaudit in model validation. Adv Water Resour 15:167–173
Aziz K, Bories SA, Combarnous MA (1973) The influence of natural convection in gas, oil and water reservoirs. J Canad Petrol Techn 12:41–47
Beck AE, Garven G, Stegena L (1989) Hydrogeological regimes and their subsurface thermal effects. Geophys Monogr vol 47. Americal Geophysical Union, Washington DC
Bredehoeft JD, Papadopulos IS (1965) Rates of vertical groundwater movement estimated from the earth's thermal profile. Water Resour Res 1:325–328
Clauser C, Villinger H (1990) Analysis of conductive and convective heat transfer in a sedimentary basin, demonstrated for the Rheingraben. Geophys J Int 100:393–414
Domenico PA, Schwartz W (1998) Physical and chemical hydrogeology, 2nd edn. Wiley, New York
Domenico PA, Palciauskas VV (1973) Theoretical analysis of forced convective heat transfer in regional ground–water flow. Geol Soc Am Bull 84:3803–3814
Forster C, Smith L (1989) The influence of groundwater flow on thermal regimes in mountainous terrain: a model study. J Geophys Res 94:9439–9451
Frumkin A, Gvirtzman H (2006) Cross-formational rising groundwater at an artesian karstic basin: the Ayalon Saline Anomaly, Israel. J Hydrol 318:316–333
Garg SK, Kassoy DR (1981) Convective heat and mass transfer in hydrothermal systems. In Rybach L, Muffler LJP (eds). Geothermal systems. Wiley, New York
Goguel J (1976) Geothermics. MacGraw-Hill, New York
Goldscheider N, Mádl-Szőmyi J, Erőss A, Schill E (2010) Review: thermal water resources in carbonate rock aquifers. Hydrogeol J 18:1303–1318
Haenel R, Rybach L, Stegena L (1988) Fundamentals of geothermics. In: Haenel R, Rybach L, Stegena L (eds) Handbook of terrestrial heat–flow density determination. Kluwer Academic Publishers, Dordrecht
Hanano M (1998) A simple model of a two-layered high temperature liquid dominated geothermal reservoir as a part of a large scale hydrothermal convection system. Transp Porous Media 33:3–27
Holzbecher EO (1998) Modeling density–driven flow in porous media. Springer Verlag, Berlin
Horton CW, Rogers FT Jr (1945) Convection currents in a porous medium. J Appl Phys 16:367–370
Jessop AM (1990) Thermal geophysics. Elsevier, Amsterdam
Kühn M, Dobert F, Gessner K (2006) Numerical investigation of the effect of heterogeneous permeability distributions on free convection in hydrothermal system at Mount Isa, Australia. Earth Planet Sc Lett 244:655–671
Lachenbruch AJ, Sass JH (1977) Heat flow in the United States and the thermal regime of the crust. In: Heacock JG (ed) The earth's crust, its nature and physical properties. American Geophysical Union, Washington DC
Lapwood ER (1948) Convection of a fluid in a porous medium. Proc Cambridge Phil Soc 44:508–521
Lu N, Ge S (1996) Effect of horizontal heat and fluid flow on the vertical temperature distribution in a semiconfining layer. Water Resour Res 32:1449–1453

Lubimova EA, Von Herzen RP, Udintsev GB (1965) On heat transfer through the ocean floor. In: Lee HL (ed) Terrestrial heat flow. Port City Press, Baltimore

Manning CE, Ingebritsen SE (1999) Permeability of the continental crust: implications of geothermal data and metamorphic systems. Rev Geophys 37:127–150

Mansure A, Reiter M (1979) A vertical groundwater movement correction for heat flow. J Geophys Res 84:3490–3496

Murphy HD (1979) Convective instabilities in vertical fractures and faults. J Geophys Res 84:6121–6130

Pasquale V, Verdoya M, Chiozzi P, Šafanda J (2000) Evidence of climate warming from underground temperatures in NW Italy. Global Planet Change 25:215–222

Pasquale V, Verdoya M, Chiozzi P (2010) Darcy velocity and Péclet number analysis from underground thermal data. Boll Geofis Teor Appl 51:361–371

Pasquale V, Verdoya M, Chiozzi P (2011a) Groundwater flow analysis using different geothermal constraints: the case study of Acqui Terme area, northwestern Italy. J Volcan Geoth Res 199:38–46

Pasquale V, Gola G, Chiozzi P, Verdoya M (2011b) Thermophysical properties of the Po basin rocks. Geophys J Int 186:69–81

Pasquale V, Chiozzi P, Verdoya M (2013) Evidence for thermal convection in the deep carbonate aquifer of the eastern sector of the Po plain, Italy. Tectonophysics 594:1–12

Pestov I (2000) Thermal convection in the great Artesian basin, Australia. Water Res Manag 14:391–403

Phillips OM (1991) Flow and reactions in permeable rocks. Cambridge University Press, Cambridge

Raffensperger JP, Vlassopoulos D (1999) The potential for free convection in sedimentary basins. Hydrogeol J 7:505–520

Reiter M (2001) Using precision temperature logs to estimate horizontal and vertical groundwater flow components. Water Resour Res 37:663–674

Schoeller H (1962) Les eaux souterraines. Masson & Cie, Paris

Smith L, Chapman DS (1983) On the thermal effects of groundwater flow 1: regional scale systems. J Geophys Res 88:593–608

Stallman RW (1963) Computation of ground–water velocity from temperature data. In: Bentall R (ed) Methods of collecting and interpreting ground–water data. Survey Water, U.S. Geol

Stallman RW (1965) Steady one-dimensional fluid flow in a semi-infinite porous medium with sinusoidal surface temperature. J Geophys Res 70:2821–2827

Swanson SK, Bahr JM (2004) Analytical and numerical models to explain steady rates of spring flow. Ground Water 42:747–759

Tóth J (1995) Hydraulic continuity in large sedimentary basins. Hydrogeol J 3:4–16

Turcotte D, Schubert GL (2002) Geodynamics–application of continuum physics to geological problems, 2nd edn. Cambridge University Press, Cambridge

Verdoya M, Chiozzi P, Pasquale V (2007) Thermal log analysis for recognition of ground surface temperature change and water movements. Clim Past 3:315–324

Verdoya M, Pasquale V, Chiozzi P (2008) Inferring hydro–geothermal parameters from advectively perturbed thermal logs. Int J Earth Sc 97:333–344

Wood JR, Hewett TA (1982) Fluid convection and mass transfer in porous sandstones—a theoretical model. Geochim Cosmochim Acta 46:1707–1713

Zhao C, Hobbs BE, Muhlhaus HB, Ord A, Lin G (2003) Convective instability 3D fluid saturated geological fault zones heated from below. Geophys J Int 155:213–220

Errata to: Geothermics

Vincenzo Pasquale, Massimo Verdoya and Paolo Chiozzi

Errata to:
V. Pasquale et al., *Geothermics*,
SpringerBriefs in Earth Sciences
DOI 10.1007/978-3-319-02511-7

Page vii, line 6	11 should be 10
Page 1, line 4	Replace "body" with "seismic"
Page 3, line 10 from bottom	Replace "of 35 km" with "35 km"
Page 13, line 22	Replace "versus the" with "versus"
Page 21, line 6	Replace "correspond" with "corresponds"
Page 31, line 4 from bottom	Replace "also" with "strongly"
Page 34, line 1	Replace "where for $a < 1$" with "where for $a < 1$ and $\theta = \cos^{-1}a$"
Page 35, Eq. (2.53)	$k_w = 0.6020 + 1.309 \times 10^{-3}\,T - 5.140 \times 10^{-6}\,T^2$ for $T > 137\,°C$
Page 36	In Figure, on the scale of thermal conductivity, replace 24 with 2
Page 55, Line 6 from bottom	1000 m and 1 mK m^{-1} at 2000 m
Page 58, Eq. (3.12) last term	$0.00012 c_K A_{K^{40}} \exp(t\ln 2/\tau_{K^{40}})$
Page 59, Line 10 from bottom	Replace "is given by" with "gives"
Page 66, Line 15	Replace "solidus" with "solidus temperature"
Page 72, Line 2 from bottom	Replace "T_{m_o} as eq. 3.43" with "T_{m_o}"
Page 114	In Figure the parameter h should be replaced by L
Page 114, line 5	Replace "k" with "κ_p"
Page 117, Line 1 from bottom	Replace "Dike, 62" with "Dyke, 93, 96"

The publisher apologizes for the inconvenience caused.

The online version of the original book can be found under
DOI 10.1007/978-3-319-02511-7.

V. Pasquale (✉) · M. Verdoya · P. Chiozzi
University of Genova, Genova, Italy
e-mail: pasquale@dipteris.unige.it

M. Verdoya
e-mail: verdoya@dipteris.unige.it

P. Chiozzi
e-mail: chiozzi_rp@tiscali.it

V. Pasquale et al., *Geothermics*, SpringerBriefs in Earth Sciences,
DOI: 10.1007/978-3-319-02511-7_6, © The Author(s) 2014

Index

V. Pasquale et al., *Geothermics*, SpringerBriefs in Earth Sciences,
DOI: 10.1007/978-3-319-02511-7, © The Author(s) 2014